A Mushroom Word Guide

Etymology, Pronunciation, and Meanings of over 1,500 Words

By Robert M Hallock, PhD

ISBN:
9781796472851

CONTENTS

ACKNOWLEDGMENTS

I would like to thank the readers of the Facebook group *Mycological Word of the Day,* which led to the writing of this book. It has been a pleasure to interact with mushroom people from beginners to some true mycological giants. Thanks to everyone who visited and engaged in the group and discussions. Lastly, I would like to thank Kris Banowetz-Hallock for the cover photo.

1. BASIC MYCOLOGICAL DEFINITIONS

adnate (ad NATE): Latin for *to grow to.* Adnate gills are broadly attached to the stem.

adnexed (ad NEXT): Latin for *bound or tied to.* Adnexed gills are narrowly attached to the stem.

aethalium (ee THAY lee um): The spore-bearing stage of a slime mold. For *Fuligo septica*, the mobile plasmodium becomes the aethalium, a cushion-shaped fruitbody capable of releasing spores.

aff.: A Latin abbreviation of *affinis* which means 'having an affinity to', and is used to express that a species is related in a compelling way. For example, *Gymnopilus aff. bellulus* was similar, but seemingly different from, *Gymnopilus bellulus*. The use of *cf* and *aff* are part of 'open nomenclature' and there are no official guidelines for their use.

anamorph (G): Literally, a form from the bottom up, taken to mean a developmental form. A fungus at an asexual stage that produces a fungus distinct from the sexual form. Also called 'imperfect fungi', 'fungi imperfecti', or just 'imperfect', although these terms are obsolete. For example, the *Xylaria cubensis* anamorph is a distinct (and sometimes more common) form of the spore-producing teleomorph.

anastomose: Anastomosing (ah NASS teh mose ing) gills are interconnected with a network of veins.

annulus (L): A ring. *Galerina annulata* (an you LAHT uh) has a prominent annulus. Remnants of the partial veil left on the stem can leave anything ranging between a faint annular zone to a prominent ring.

apothecium (G): A storehouse, pronounced (ap oh THEE see um). A bowl or cup-shaped fruitbody, with the hymenium on the upper concave surface like in *Helvella*. The morel has a number of apothecia fused together. The word is related to 'apothecary', referring to either a storehouse (typically a pharmacy) or a storekeeper (typically a pharmacist).

ascocarp (G): Bag / fruit, pronounced (ass co carp). A sporocarp (fruitbody) that produces asci.

ascomycete (G): Bag / mushroom, pronounced (ass co MY seat). A taxonomic group of fungi that produce spores in asci. This group includes morels, truffles, ergot, cup-fungi, *Xylaria*, *Cordyceps*, and many others. What kind of mushroom-bag do you have?

ascus / asci (G): A leather bag or bladder. Asci (ass kie) is plural and ascus (ass cuss) is singular. An ascus is a sack that typically develops eight spores.

basal mycelium (BAZE ul - my SEAL ee um): Latin for 'base mycelium'. Basal mycelium is the mycelium present at the base of the stipe, and it can be very important in identification. For example, *Laccaria bicolor* has purple basal mycelium, and it is only purple for a few minutes or so after it is picked. Many collections of *Laccaria* may be incorrectly identified because the basal mycelia is not noticed.

basidiomycete (buh sid ee oh MY seat): Latin for 'a mushroom with a small pedestal'. Basidiomycetes are mushrooms with basidia, and they include gilled mushrooms, puffballs, polypores, pored-mushrooms, toothed-fungi, jelly fungi, and more. Contrast with the the ascomycetes.

basidium (buh SID e um): Latin for 'a small pedestal'. The basidium is located on the hymenium of a mushroom (at least on the basidiomycetes). The basidium produces and supports the spores, and a typical basidium holds four spores (exceptions exist and includes *Amanita bisporigera* which has two spores on each basidium). Basidium is singular and basidia is plural. If you have trouble remembering which is singular and which is plural, you might know that bacteria is plural and bacterium is singular - these words have the same suffixes.

beaked peristome (PEAR ih stome): The peristome on some earthstars is shaped like a beak.

caulocystidia (CAU lo sis tid ee uh): Caulocystidia are cystidia that are on the stem.

cf: A Latin abbreviation that is short for *conferre* and means 'compare to'. The *cf* is typically placed between the genus name and specific epithet. It is commonly used when naming a species to indicate a low confidence in the name or that the specimen does not exactly fit the proposed species name.

chambers: If you take a cross section of a puffball, you might see irregular chambers in the gleba, or a chambered subgleba.

cheilocystidia (KEY lo sis tid ee uh): Greek for 'margin'. Cheilocystidia are a special type of cystidia that are on the edges of the gills.

clavate (L): A club (CLAH vate or CLAY vate). This is an adjective frequently used to describe a club-shaped stem.

collar: In many earthstars, the bottom of the gleba has a collar that spreads out over the pedicel. *Geastrum fornicatum* and *G. bryantii*, and other earthstars have a prominent collar. In some, the collar is somewhat flaring and is seen partially hanging over the stalk.

connivent, converging fibrils, or spines: Fibrils of a puffball whose tips converge together are called connivent or converging. Typically, the tips of up to four fibrils converge together. Next time you see some spines, check them out under a loupe, you might be surprised at what you can see! *Lycoperdon echinatum* (= *L. americanum*) has convergent tips.

crenate (L): Notched, scalloped (Pronounced CREE nate). This is a common adjective to describe the scalloped appearance of the margin of the pileus. 'Crenulate' means finely crenate.

cystidia (sis TID e uh): Greek for 'bladder' or 'pouch'. Cystidia are sterile cells found near basidia. They are similar in appearance to basidia, but they are often larger, and they have no spores or sterigmata. There are numerous types of cystidia, and they have particular shapes that are important in identification.

deciduous (dee SIDGE you us) **warts:** Warts on the exoperidium of many puffball species can be scraped off (or otherwise fall off with age), and these are called deciduous warts. They are common in *Lycoperdon perlatum*, which also has textbook pyramidal warts.

decurrent (dee CURR ent): Decurrent gills run down the stem. Examples include *Pleurotus ostreatus* (the oyster mushroom) and various *Clitocybe* species.

desiccate: To make thoroughly dry (pronounced DESS ih cate). Used in mycology to describe the process of drying specimens for an herbarium. A desiccator (food dehydrator) is used to dry fungal specimens, and the desiccated specimen must be kept dry while in storage. To accomplish this, a desiccant, like water-absorbing silica beads, may be added to ensure proper long-term storage.

emarginate (ee MAR jin it): The gills are notched at the attachment to the stem (and do not run down the stem like they do with a sinuate attachment).

endoperidium (EN doe pear ID ee um): From the Greek word *endo*, meaning 'inside'. When the peridium has more than one layer, like in the puffball, the endoperidium is the inner-most layer.

epithet (EP ih thet): The epithet is the second part of the scientific name. You will often encounter the phrase, 'the specific epithet', which generally just means the second half of the species name (but doesn't it sound much nerdier?). For example, the specific epithet of *Morchella esculenta* is *esculenta.*

eponymous: Something named after a person. There are many eponymous (ee PON ih mus) mushroom species that are named after the person who discovered the mushroom, or named in honor of someone.

equal: The same. This is frequently used when describing a mushroom. A mushroom has equal gills if they are all the same length. A mushroom's stem is equal if it is the same thickness throughout the whole stem.

exoperidium (ex oh pear ID ee um): From the Greek word *exo*, meaning 'outside'. When the peridium has more than one layer, like in the puffball, the outer-most layer is the exoperidium. This outermost layer is also typically ornamented, like having pyramidal warts.

exudate (EX you date): From the Latin *exudare*, meaning 'to sweat'. In the context of mycology, it is a fluid that oozes from a surface of a mushroom. There is also mycelial exudate, a liquid forming on the surface of mycelium in some species. *Hydnellum peckii* has an exudate, as do many others.

fibrillose peristome: A fibrillose peristome is made of thin fiber-like filaments. *Geastrum triplex* is a good example that has a fibrillose peristome. Check them out under a small magnifying glass to see these structures up close.

fibrils or spines: Some puffballs have fibrils on the exoperidium instead of warts. Connivent and tapering fibrils are also defined here. *Lycoperdon echinatum* (= *L. americanum*) has fibrils.

flaring annulus: The annulus on the stem of a mushroom can take several forms. A flaring annulus spreads out over the stem like a miniature skirt or umbrella. Look for a flaring annulus on the deadly *Amanita virosa* and *A. bisporigera* (at least in mature specimens).

flavescent: An adjective to describe a change to yellow. For example, *Agaricus xanthodermis* turns yellow when bruised or cut, and when Potassium Hydroxide is applied to it.

free gills: Free gills are not connected to the stem. *Amanita* species are most commonly associated with free gills, but there are many other species with this type of attachment, like *Agaricus*, *Chlorophyllum*, and *Lepiota.*

gleba (GLEE buh): Latin for *clod*, which is a lump of earth. A spore-bearing mass of tissue is called the gleba. For a puffball, this is the inside mass of spores, but in a stinkhorn, the gleba is the spore bearing mass on the head of the fruitbody (or sporocarp, which is also defined here). The gleba of a young edible *Lycoperdon* is uniformly white. With age, it will turn yellow, and then greenish, and will finally be brown with mature spores when the puffball is older (the puffball will also begin to develop a distinct odor when the spores turn yellow, which is another indication that it should no longer be consumed). The next time someone asks you if a puffball is edible, you can ask them what color the gleba is.

guttate (GUT tate): Guttate means that the fungus is exuding droplets of water. For fungi, many species give off droplets of water during periods of rapid growth, and the exudate is a byproduct of the metabolism (Parmasto & Voitk, 2010). *Hydnellum peckii* has red droplets, which are just water droplets with a red pigmented chemical called terphenylquinone.

guttation drops (gut TAY tion drops): Rather than the adjective to describe a mushroom that exudes drops of liquid (guttate), a guttation drop is the noun to describe the actual droplets.

holotype: Holo is Greek for 'whole' or 'entire'. The holotype of a species is the physical specimen of the species that was originally described when it was named. Although the holotype may be a single specimen, it may also be a collection of specimens that was used to characterize the species in the original publication, and they are typically found in herbaria. Species with the epithet *hololeuca* are entirely white, and species with the epithet *holophaea* are entirely dark. Related words include 'hologram' (whole image) and 'holocaust' (whole burn).

hymenium (hi MEN e um): Greek for 'a membrane'. The hymenium is the spore-bearing surface of a mushroom. This includes the gills for the gilled mushrooms, the ridges of a chanterelle, the tubes for pored mushrooms, the inside of the fruitbody for puffballs, the spines of toothed-fungi, the surface of the cap for morels, and the inside of the cup for cup fungi.

hypha (HIGH fah): Greek for 'web'. Hyphae are cells that collectively make up mycelium. Hypha is singular and hyphae is plural.

indexfungorum.org: A website to look up information about the names of mushroom species. From the main page, you can select 'Search Indexfungorum' on the right and then enter various terms in the search box. For example, you can click 'Epithet', and then type in 'emetica', and then all species that utilize 'emetica' in the specific epithet will come up. Or, you can click 'Genus', and search 'Russula', and all Russula species will come up. You can also check to see if you are using the most current name for a fungus. For example, if you are using an old mushroom book, you might come across '*Pycnoporus cinnabarinus*'. Type that in the 'Name' box, and you will see the current name is '*Trametes cinnabarina*'. Search away!

inequihymeniiferous / inaequihymeniferous (in eh kwee hi men IF er us): Literally, 'not / equal / spore bearing surface / producing / like'. This term refers to the uneven maturing of spores on the hymenium (spore bearing surface), which causes a mottled appearance on the surface of the gills. It is a common feature to see in a few species, most notably in *Panaeolus* and *Lacrymaria*. This word was coined in 1909 by Arthur Henry Reginald Buller, a British mycologist. Note that the British spelling is with an 'ae' in the beginning, while American English spelling would convert this to just an 'e'. Further, some sources spell the word with a single 'i', and I have no explanation for this alteration other than that SpellCheck doesn't know what this word is.

lacerate peristome: A lacerate peristome has irregularly torn edges (is lacerated). Many *Geastrum* species develop this feature with age.

lamella (luh MEL uh): Latin for 'a thin plate'. This is the scientific word for 'gills'. Lamellae (luh MEL ay) is plural and lamella is singular.

lamellulae (luh MEL you lay): Latin for 'a little thin plate'. Lamellulae are literally short gills - they do not go all the way between the edge of the cap and the stem.

mycelium (my SEAL ee um): Greek for 'a fungus'. Mycelium is a mass of hyphae – whether it is the vegetative part of the fungus or a white mass of hyphae in the substrate from which the mushroom grows.

mycoheterotroph (literally fungus-different-feeding) describes a relationship between plants and fungi. One of the best known examples is the Indian Pipe (*Monotropa unifora*), which is a parasite on the mycelium of Russula and Lactarius.

nom. prov.: A Latin abbreviation of *nomen provisorium* which means 'provisional name'. It is used when a species lacks a formal peer-reviewed publication. For example, "*Amanita amerirubescens* Tulloss nom. prov." is described, accepted in most circles, but remains (as of this writing, I believe), unpublished.

notched gills: Notched gills is a broad term that encompasses both sinuate and emarginate gills. The next time someone says a mushroom has notched gills, you can clarify whether they are emarginate or sinuate (both are defined here).

ostiole (OS tee ole): The pore at the top of a mature puffball where spores are released is called the ostiole.

ozonium (oh ZOE nee um): A carpet of rhizomorphs at the base of a fruitbody, which are typically orange. *Coprinellus domesticus* has been called the 'firerug inkcap', and Michael Kuo calls it 'the retro inky' due to the ozonium.

partial veil: A partial veil is a thin membrane that covers the developing gills or pores of some mushrooms. It goes between the stem and edge of the cap. Sometimes you will see remnants of the partial veil hanging off the edge of the cap. Portions of the partial veil can also be left on the stem in the form of a pendant annulus (also described here).

patches: Remnants of the universal veil on the top of the cap appear as warts in some species, like in *Amanita muscaria.* Sometimes, however, the remnants are large pieces of universal veil called patches, like in *A. velosa.* Although patches or warts can be a useful identifying characteristic, it is important to note that they can sometimes wash off the cap with a good rainfall.

pedicel: A short stalk upon which the gleba is elevated in many earthstars.

pendant annulus: The annulus on the stem of a mushroom can take several forms. A pendant annulus hangs down on the stem but closely hugs it like a tight-fitting skirt. Look for a pendant annulus on *Amanita muscaria*, at least in mature specimens.

peridiole: A basidia-containing chamber. In *Pisolithus*, these are inside the fruit body. In *Cyathus*, the peridioles (pare ID ee oles) are the 'eggs' inside the cup. It is from the Greek *peridium*, meaning 'small wallet'.

peridium (pear ID ee um): From the Greek word *perideris*, meaning 'wallet'. The outer layers/layers of the fruitbody of a gasteromycete is called the peridium. The peridium of the puffball is like our skin: it keeps the inside in, the outside out, and it has several layers.

peristome (PEAR ih stome): From the Greek words *peri* and *stoma*, meaning 'around the mouth'. This is the area around the opening where spores are released in the gasteromycetes. *Calostoma* species have rather pronounced peristomes.

pileipellis (pil EY ih pell is): Literally, 'cap-skin', this is the outer membrane of the pileus. If this outer layer is peelable, it is called the pellicle.

pileosetae (L): Literally, 'hairs on cap', pileosetae are single-celled cystidia present on a number of species, including *Parasola conopilus* and those in Mycena section Longisetae.

plasmodium (plaz MOE dee um): The mass of cells of the slime mold that is able to move around in the environment. For example, the yellow plasmodium of the dog-vomit slime mold can often be found on mulch and gardens, and cruises along at around at 1 mm per hour.

pseudocolumella (SUE doe col oh MEL luh): Sterile tissue surrounded by gleba. It appears as a false column of sterile tissue.

radicating: Rooting, extending deep into the substrate (Pronounced RAD ih cay ting). Used to describe the base of the mushroom stalk that extends into the substrate, also called a tap root. The radicating stem is distinguished from a tapering base, which comes to a point but only superficially extends into the substrate.

ray: Rays are essentially the legs of an earthstar. How many rays does your earthstar have?

recurved: Describes a margin that flares upward, like in *Peziza repanda*, the 'recurved cup'.

recurved rays: When the rays of an earthstar are convex such that they elevate the fruitbody, they are called recurved rays.

resupinate (reh SUE pin ate): A flat (effused) fruitbody with the hymenium facing upwards or just at the periphery. From the Latin *supinus* or *resupinare*, which mean 'lying on the back'. You may be familiar with a supine positon when an animal or person is flat on their back, like when a dog wants its belly rubbed.

rhizomorph (RYE zoe morph): From the Greek *rhiz*, meaning 'root'. A rhizomorph is a root-like aggregation of hyphae at the base of a mushroom. Individual strings of hyphae are called rhizomorphs, and a collection of rhizomorphs is called mycelium. You can see rhizomorphs at the base of many *Mycena* species. Additionally, the honey mushroom, *Armillaria*, is also called the 'black shoestring mushroom', and gets its common name because of the black rhizomorphs. The presence and color of rhizomorphs are sometimes useful in identification, so the careful extraction of the base of the mushroom is critical to a good collection and reliable identification.

rufescent: An adjective to describe a change to red, which is typically fast. For example, *Agaricus fuscofibrillosus* stains red when cut.

rugose (L): Wrinkled, corrugated. 'Rugose' (ROO gose) is a common adjective to denote something that is wrinkled, while 'rugulose' (ROO gyou lowse) means 'finely wrinkled'. In addition to adjectives used in descriptions, these two words are used in over 250 species names for mushrooms.

sensu auct.: is short for '*sensu auctorum*', Latin for 'in the sense of the author'. It can be followed with a particular author's name to read "In the sense of author X". You may also see '*sensu auct. Amer.*', meaning "In the sense of American authors". This is typically used after the name of a

European species that we realize is probably not the same species in the Americas.

sensu lato (L): In the broad sense. For example, "*Amanita muscaria sensu lato* has a broad distribution" would mean that *A. muscaria*-related species are found throughout the world. This phrase is commonly used in the mushroom identification boards.

sensu stricto (L): In the narrow sense. For example, you might see "This is not *Auricularia auricular-judae sensu stricto*, it is probably *Auricularia americana* since it was found in New York" in a mushroom identification board.

serrated: Some mushrooms have gills that are serrated like a knife. *Lentinellus*, *Neolentinus* and *Hebeloma* have specimens with distinctly serrated gills.

sessile (L): Fixed in place. Sessile (SES ill or SES isle) fungi lack a stem, as with puffballs, earth stars, and some other fungi. In zoology, sessile animals like barnacles are permanently attached to a substrate.

shallow pits: Shallow pits are left when deciduous warts are removed or fall off.

sheathing annulus: The annulus on the stem of a mushroom can take several forms. A sheathing annulus opens up towards the cap of the mushroom. Look for a sheathing annulus on the mature *Agaricus bitorquis*, among others.

short gills, sub-gills. Members of *Russula* section Compactae have short gills / sub-gills, which are gills that do not go all the way from the stipe to the edge of the pileus. Many other species share this characteristic. Synonymous with, although less formal than, lamellulae.

sinuate (SIN you ate): Latin for 'curve'. Gills with a sinuate attachment are notched and then briefly run down the stem.

sp. nov.: A Latin abbreviation of *species nova*, which means 'new name'. It is used when a species is published for the first time.

spore (s PORE): Greek for 'a seed'. Taken literally, spores are the seeds of fungi (and the next time someone says that spores are not seeds, you can inform that that the word comes from the Greek word meaning seed). Spores are microscopic, and many are around 5 micrometers in diameter (1/200th of a millimeter). Puffballs can produce trillions of spores. When you take a spore print, the deposit of spores will be a particular color, and the color is useful for identification.

spore dusting: A dusting of spores that you can sometimes find in the field if you look carefully at whatever is growing under a mature mushroom cap. Frequently, you can just look under the most mature fruitbody at the top of an underlying cap and find the spore dusting. They are commonly found in *Armillaria* and *Flammulina* species, which both grow in clusters and have white spore prints. Sometimes the spore dusting can be on the ground (frequently seen with *Ganoderma*), on the stem of the mushroom itself, or even on spider webs that are under the mushroom. Crystal Davidson took a great picture of a pink-tinged spider web that was under a *Rhodotus palmatus.*

sporocarp (SPORE oh carp): This is another word for the fruiting body of a mushroom.

sterigma (ster IG muh): Greek for 'support, a prop'. A sterigma is a protrusion from a basidium that holds a spore. Sterigmata is plural and sterigma is singular.

stipitipellis (stip it ih PELL is): Literally, 'stalk-skin', this is the outer layer of the stipe. In some older *Armillaria* specimens, the tough outer layer of the stipe (stipitipellis) can be readily peeled off.

subdecurrent (SUB dee CUR ent): Subdecurrent gills run down the stem, but not as much as in a decurrent attachment. *Chroogomphus* species are good examples of mushrooms with subdecurrent gills.

subgleba: Tissue that supports the gleba. In *Lycoperdon perlatum*, the subgleba is the stalk-like base of the puffball.

sulcate (SULL cate) peristome (PEAR ih stome): From the Latin *sulcus*, meaning 'a groove or furrow'. A sulcate peristome has grooves. *Geastrum pectinatum* has a sulcate peristome.

tapering: Coming to a point. Used to describe the base of the mushroom stalk that comes to a point. This is distinguished from a radicating base, which extends much deeper into the substrate.

tapering fibrils, spines: Fibrils whose tips do not join together with other fibrils are called tapering fibrils (or spines). *Lycoperdon foetidum* has tapering spines.

teleomorph (G): End, far from, taken to mean the final developmental form. A fungus at this stage will produce spores when mature, and is distinguished from an anamorph which will not produce spores. Also called 'perfect fungi' or 'fungi perfecti', although these terms are obsolete because they are no more perfect than anamorph fungi. *Xylaria cubensis* has a spore-producing teleomorph form as well as a common non-spore producing anamorph form. Related words include 'television' (vision from a distance) and 'telencephalon' (the portion of the brain farthest from the brainstem).

toadstool: A toadstool can be broadly defined as a mushroom with a cap, gills, and stem. However, more commonly, it is used to describe a poisonous mushroom. From John Steinbeck's *To a God Unknown*: "There were millions of mushrooms in the field, and puff-balls and toadstools too. The children brought buckets of mushrooms in, which Rama fried in a pan containing a silver spoon to test them for poison. She said that silver would turn black if a toadstool was present." (p. 83). Here, Steinbeck clearly defines a toadstool as a poisonous mushroom. Mushroom authors commonly made the same distinction in the early 20th century. Book titles from the era include 'Mushrooms and Toadstools' by Gussow and Odell in 1927. Gibson takes the position that a toadstool is a mushroom where the edibility is unknown. To that end, he cleverly named his book 'Our Edible Toadstools and Mushrooms and How to Distinguish Them' (W Hamilton Gibson, 1895).

trama (TRAY ma): Latin for 'woof' or 'weft' (woven fabric). Trama is the inner layer of interwoven hyphae, usually of the gills. I am unsure why it is spelled 'trauma' on several online sites.

type species: A type species is the species to which a genus (or subgenus) is permanently linked: it is the first species described in a genus (or subgenus).

universal veil: A universal veil is a membrane that fully ensheathes some species of gilled mushrooms when they are immature. When the mushroom matures, remnants of the universal veil are sometimes visible on the surface of the cap as warts (think of the warts on the cap of *Amanita muscaria*).

volva (L): A covering. Used in *Amanita volvata* (vole VAH tuh). Remnants of the universal veil at the base of the stem is called a *volva*, and it is present on many gilled mushrooms, particularly Amanitas. Remnants of the universal veil remain and can be seen as the cup-like sac that we call the volva in a number of species.

2. MYCOLOGISTS & EPONYMOUS EPITHETS

Many of the mycologists treated here also have species named after them and are included in the main glossary of mycological terms. This list represents a comprehensive list of names treated throughout the book:

Alessio, Carlo Luciano
Allen, Alissa
Allen, John W
Archer, William
Arora, David
Arrhenius, Johan Peter
Badham, Charles David
Ballou, W. H.
Banker, Howard James
Banning, Mary Elizabeth
Barrows, Charles "Chuck"
Battarra, Giovanni Antonia
Beauverie, Jean
Berkeley, Cecilia Emma
Berkeley, Miles J
Birnbaum:
Blakesley, Albert Francis
Bloxam, Andrew
Boone, William Judson
Both, Ernst E
Bresadola, Giacopo
Buchwald, Niels Fabritius
Cajander, Aimo Kaarlo
de Candolle, Augustin
Claussen, Peter
Cogniaux, Alfred
Coker, William Chambers
Curtis, Moses Ashley
Darwin, Charles
Dudley, William Russel
Flett, J. B.
Franchet, Adrien René
Fries, Elias Magnus
Frost, Charles Christopher
Fuckel, Karl
Gautier, Joseph
Gerard, William Ruggles
Gilbertson, Robert Lee
Goetz, Christel & Donald
Guépin, Jean Baptiste
Güssow, Hans T
Guzman, Gaston
Heim, Roger Jean
Hesler, Lexemuel Ray
von Hochstetter, Ferdinand
Hodge, Kathie
von Hohenbühel, Ludwig
Imler, Louis Philip
Jackson, Henry
Jahn, Hermann Theodor
Junius, Hadrianus
Konrad, Paul
Korf, Richard "Dick" Paul
Kretzschmar, Eduard

Krieger, Louis
Kuehner, Calvin C
Kühner, Robert
Lasch, Wilhelm Gottfried
Lea, Thomas Gibson
Lecomte, Paul Henri
Le Gal, Marcelle Louise
Lenz, Harald Othmar
Le Rat, Auguste-Joseph
Lloyd, Curtis Gates
Lyon, H. L.
Mao, Lan
Michener, Ezra
Montagne, Camille
Morgan, Andrew Price
Morris, George Edward
Murray, Dennis
Murrill, William Alphonso
Nannfeldt, John Axel
Oudemans, Cornelius
Paden, John Wilburn
Peck, Charles Horton
Perceval, Cecil H. Spencer
Pouzar, Zdenek
Puccini, Tommaso/Tomaso
Quelet, Dr. Lucien
Ravenel, Henry William
Ricken, Adalbert
Ristich, Samuel S
Rodman, James
Romagnesi, Henri
Russell, John Lewis
de Schweinitz, Lewis
Smith, Alexander Hanchett
Smith, Worthington George
Snyder, Leon Carlton
Sowerby, James
Sprague, Charles James
Squarepants, Spongebob
Stahel, Gerold
Stevenson, Rev. John
Sullivant, William Starling
Sumstine, David Ross
Thaxter, Roland
Thiers, Harry D.
Trogia, Jacob Gabriel
Tulasne, Louise
Tulloss, Rodham (Rod) E.
Wakefield, Elsie M
Wasson, R Gordon
Weaver, Evelyn M "Peg"
Whetsone, Mary S
Yates, Harry Stanley
Zeller, Sanford Myron
Zollinger, Heinrich

3. CHEMICAL TESTS

Ammonium Hydroxide / Ammonia (NH4OH): A few drops are typically applied to the outer and cut surfaces of a bolete to test for color changes. 'North American Boletes' by Bessette, Roody, and Bessette have a subsection called 'Macrochemical tests' for each species they describe in the book. Many entries include information on NH4OH. *Boletus edulis*, for example, 'pileipellis stains orange with the application of KOH or NH4OH'. Frequently, a reaction may start out one color then rapidly change to another. For example, it might 'flash green, then orange'.

The solution: Mycology typically uses a 5-10% ammonia solution, which is good because household ammonia contains the same amount. You should buy the clear ammonia, not the yellow one that has an added citrus odor.

amyloid (AM il oid): Comes from the Greek *amyl*, meaning *starch*. Spores that stain blue (bluish-black) with Iodine (including Melzer's reaction) are amyloid. The color change results from the presence of starch (thus the name amyloid). As far as the spore goes, the spore wall contains starch. However, paper contains starch, too, so you cannot do the chemical test on a spore print that is on paper, nor can you scrape a pile of spores from a spore print on paper and transfer it to glass. You can take a spore print directly on glass, scrape the spores into a pile, and then add a drop of solution to test for the color change.

Congo red: A chemical that comes from an evergreen tree in South America. This is a stain for cell walls (walls of hyphae), and is commonly used to examine clamp connections. Emmett (2003) describes how to use this chemical in fungal microscopy.

The Solution: Often a solution in water (1%) or ammonia (10%). Aqueous solutions are less preferred, but readily available on Amazon.

dextrinoid: Spores (or tissue) that change reddish-brown (often described as chestnut brown) in response to Melzer's solution are dextrinoid.

guiac / guaiac: If you open 'Fungi of Switzerland Volume 6 Russulaceae', you'll find that the 'chemical reactions' section lists results for 'guiac' on every entry for the Russulas (but not *Lactarius*). It was first used in mycology in 1910, and this chemical is an important diagnostic for identification of Russulas. A positive reaction to a Russula will stain the tissue (typically tested on the stipe) some form of blue or green. Some descriptions in Fungi of Switzerland include 'blue-green', 'dark green', 'dark olive green', 'light green', and 'pale green'. If there is no reaction within 4 minutes, the test is negative.

The Solution: The oil, resin, and gum from the wood comprise Guaiac (Tortelli, 2004).

inamyloid (IN am il lloyd) / **non-amyloid**: Spores (or tissue) that do not change color in response to Iodine (including Melzer's solution) are inamyloid. These spores are the same color before and after Melzer's reagent.

Iron Sulfate (FeSO4): A few drops are typically applied to the outer and cut surfaces of a bolete to test for color changes. Again, 'North American Boletes' by Bessette, Roody, and Bessette have a subsection that details color changes with FeSO4 in numerous species. For example, with *Boletus badius*, 'context stains dull blue-green with application of FeSO4'. This chemical is also very important in differentiating species within the genus Russula.

The solution: Mycologists typically use a 10% solution. You can buy a small bottle of it on different supply sites. You can also buy Iron Sulfate by the pound at garden centers (and on Amazon) as FeSO4-7H2O for less than $10 a pound. You will have to put it in solution yourself (one person who bought it on Amazon made a comment that they use a 1:5 solution of FeSO4-7H20 to water) for mycology.

Lugol's solution: This was named after the French physician Jean Lugol (1786 -1851). He first made this solution in 1829 to treat tuberculosis (it apparently did not work). Now, the solution is used to disinfect water and to test for starch in mushroom spores (to test for amyloid spores). This solution does not give the same results as Melzer's.

The Solution: The solution is typically 1%, 2%, or 5% iodine. A 2% solution will contain 2 g of iodine and 4 g of potassium iodide with distilled water to 100 mL. Or, you can just pick up a couple vials on Amazon for $15.

Meixner's test: This is used to test for the presence of amatoxins (poisonous compounds in deadly species like *Amanita* or *Galerina*). This test requires a cheap newspaper (or other high lignin paper) and hydrochloric acid (HCl). I will describe a standard experiment here. Three small circles are drawn onto the newspaper and labeled A, B, and C. For A, juice from a suspected amatoxin-containing mushroom (*Amanita phalloides*, for example) is expressed into the circle. For B, juice from a mushroom not suspected to contain amatoxin is put within the circle. The newspaper is allowed to dry (the newspaper cannot be subjected to heat to facilitate the drying process, as this can result in false positives). Next, a few drops of concentrated HCl are added to circles A and B. Additionally, the same amount of HCl is added to circle C. If a circles turn blue, then the reaction for amatoxins is positive. Circle C is a negative control, so if this circle turns blue, then there was a problem with the test (for example, drying the newspaper on a hot day in sunlight can apparently yield false positives). Importantly, there are many non-amatoxin containing species (including psilocybin-containing mushrooms) that will result in a blue color in this test. However, there are apparently no species that *look like* amanita-containing species that will yield a positive result with this test. You might want to choose a couple species for your control tests. See Fiedziukiewicz (2013).

The solution: You might be wondering, 'where do I get me some concentrated HCl'? You can approach your friend who works in a chemistry lab, and ask for a small little vial of their 37% HCl, or you can buy some muriatic acid on Amazon. Muriatic acid is a common additive in pools, and it is actually just an older name for HCL (After all, who would want to put concentrated hydrochloric acid in their pool?). You can get it at any place that sells pools or pool equipment, or you can just get it on Amazon. Dealing with a gallon of concentrated HCl is no joke. You need to take proper precautions - you should have proper personal protective equipment, an adequate workspace, a safe place to store it, and no kids or pets around. You should keep the Material Safety Data Sheet (MSDS) with it too.

Melzer's reagent: A few drops are typically applied to mushroom spores (or other fungal tissue) to look for color changes. Melzer's reagent will test whether spores are amyloid, non-amyloid, or dextrinoid (these terms are all defined here).

The solution: Melzer's reagent contains Potassium iodide (~3%), iodine (~1%), water (~48%), and chloral hydrate (~48%). A solution with similar proportions of chemicals is Lugol's reagent. Chloral hydrate is the only chemical here that is hard to get a hold of, and it is a controlled substance. It is a hypnotic and works centrally through GABA agonism. It was introduced for use in people in the mid 1800's, but was used as a date-rape drug in more modern times. Chloral hydrate is a Schedule IV drug in the United States. However, Melzer's Reagent is readily available in the UK and Europe. Leonard (2006) reports side-by-side comparisons between Melzer's and non-chloral hydrate iodine solutions and finds differing results between the chemicals. However, there is a proprietary solution called Visikol (search on the internet) which is made by a company that claims it is an excellent substitute for Melzer's. I have not used it (probably because it is $70 plus shipping for 15 mL).

phenol / carbolic acid / C6H5OH: 'Fungi of Switzerland Volume 6 Russulaceae' has the color reaction to phenol listed for every entry of Russulas (but not *Lactarius*). A positive reaction to a *Russula* will stain the tissue some form of red or brown. Some descriptions in Fungi of

Switzerland include 'wine-reddish', 'wine-brown', 'light red-brown', 'pink-brown', and 'light pink'.

The Solution: It is typically a 2% or 3% solution, and you can currently buy a bottle of 90% phenol / 10% water on Ebay for $25.

Potassium hydroxide (KOH): A few drops are typically applied to the surface of a mushroom to test for color changes. For example, *Agaricus xanthodermis* will instantly turn a bright yellow with application of this chemical.

The solution: The most frequently used concentration is a 3% solution. To make a 3% solution, you mix 3 grams of KOH into 100 milliliters of distilled water. Unfortunately, I don't know where people who do not have access to a university chemistry lab can get it. I have heard that liquid Drano (which has Sodium Hydroxide in it instead of Potassium hydroxide) may act as a substitute. One needs to do side-by-side comparisons with 3% KOH to establish the reliability of Drano.

Schaeffer (Schäffer) reaction: This test is named after Julius Schäffer (1882-1944), a mycologist who devised this test that is useful in the identification of *Agaricus* species. Notably, Wikipedia claims that he is the only modern mycologist to die of mushroom poisoning. Results for this test are commonly reported with descriptions of *Agaricus* species. For example, results to this test are given in 'The Agaricales (Gilled Fungi) of California 6. Agaricaceae' by Richard W Kerrigan.

The Solution: Kerrigan reports the results as "Anniline X HNO3", and makes no explicit mention of the 'Schaffer reaction', except in the summary Table of chemical results. For this test, you make two intersecting lines of the aqueous chemicals. The first is a solution is nitric acid (HNO3), and the second is aniline. The test is positive if there is a bright orange color reaction at the intersection of the lines.

Sulfovanillin: A chemical used in Russula identification. A positive reaction will often entail cystidia turning gray or black in this solution. For example, Bart Buyck's key to Russula subsection Virescentinae contains many details about reactions to sulfovanillin.

The Solution: It is made from 98% sulfuric acid (H2SO4) and vanillin crystals. You can apply the solution in one of two ways. First, you can make up a solution of 65% H2SO4 and 35% vanillin crystals. This solution apparently will not last long, so do not make up too much. Second, you can coverslip some material with water. Put some vanillin crystals on one side of the coverslip, then add a drop of H2SOS4 and draw the solution to the other side of the coverslip. Working with sulfuric acid is not for beginners or clumsy people. Use proper personal protective equipment and be safe.

Ultraviolet light (Black light): Although this is not a chemical test, it still fits best in this section. Want to be the life of the party at the next foray? All you need to do is wait until the sun sets, the wine bottles are uncorked, and break out your black light. Head away from the party room and head over to the specimen room. Say hi to the identifiers (and ask if they need anything), then take out your black light and slowly move it over the specimens. Let the identifiers know if you see anything light up under the ultraviolet light. A black light can be purchased for $15 on Amazon, and it provides loads of fun and entertainment at forays and home. What species fluoresce under UV light and which do not? Take a mushroom, bring it in a dark room, and turn on your black light. You'll see some glow on some *Mycena* and *Dermocybe* species, too. Experiment, as you can make contributions to mycology. Also, you can set up a really awesome display for a mushroom fair using a black light display with glowing mushrooms (you would put them in a mostly-enclosed box with a black light inside).

4. MUSHROOM TOXINS

amatoxin (AM uh tox in): Amatoxins are found in a number of species, including *Amanita bisporigera*, the 'Destroying Angel', *Amanita phalloides*, the 'Death cap', *Galerina marginata*, the 'Deadly Galerina', and *Lepiota brunneoincarnata*, the 'Deadly Dapperling'. The toxin is heat resistant, and interferes with messenger RNA, and thus protein synthesis. Toxicity can result in short-term effects, a 'false-recovery', and then organ failure a day or two later. See 'Silibinin' for a new treatment for amatoxin poisoning.

coprine (CO prine): Coprine is a toxin in *Coprinopsis atramentaria*, the 'alcohol inky' cap, and related mushrooms. Once ingested, coprine is metabolized into 1-aminocyclopropanol, which inhibits the breakdown of alcohol in the liver. Normally, ethanol is broken down into acetaldehyde by an enzyme called alcohol dehydrogenase (ADH). Then, another enzyme, aldehyde dehydrogenase (ALDH) breaks that down into acetate which is excreted through urine via the kidneys. One of the oldest treatments of alcoholism is Antabuse (disulfiram), a medication that inhibits ALDH and results in a buildup of acetaldehyde. This buildup results in nausea, vomiting, tingling of the extremities, dizziness, a flushed face, and increased heartrate about an hour after alcohol consumption. The metabolic byproduct of coprine, 1-aminocyclopropanol, also inhibits ALDH and can result in the same symptoms as Antabuse. And that is why you should avoid alcohol with the alcohol inky caps.

ergot (air GOT) alkaloids: Ergot is a common name for a group of ascomycetes in the genus *Claviceps* (*C. purpurea, fusiformis, sorghi*, and others). These contain Ergot alkaloids (e.g. ergotamine and ergoline) that can cause hallucinations and/or gangrene. Interestingly, it is the vasoconstrictive properties of these chemicals that can lead to gangrene that were being examined at Sandoz Laboratories by Albert Hoffmann to treat migraines. One of the chemicals Hofmann synthesized was lysergic acid diethylamide-25, the 25th form of the chemical he made. On April 19th, 1943, he accidentally consumed some and had a full-blown hallucinogenic experience on his bicycle ride home. This day is now celebrated by some as 'bicycle day'. The hallucinogenic properties of ergot and LSD (as well as psilocybin mushrooms and peyote) are due to their similarity to the neurotransmitter serotonin. However, the *Claviceps* species are very dangerous to use because one cannot predict if they will result in hallucinations or gangrene, or worse yet, both.

gyromitrin (gy rho MY trin): The toxin is found in *Gryomitra esculenta, ambigua, infula*, and other related species. Gyromitrin is hydrolyzed into monomethylhydrazine (MMH), a chemical found in rocket fuel. Exposure to MMH can cause gastrointestinal symptoms including headaches and nausea, as well neurological symptoms including delirium, dizziness, and passing out. Parboiling these mushrooms, as is common amongst those who consume them, can result in these symptoms if the steam (which contains MMH) is accidentally inhaled. If ingested, the chemical can cause liver damage.

Ibotenic acid (eye bo TEN ic): One of the two active chemicals found in *Amanita muscaria* and *A. pantherina* (as well as other species). Ibotenic acid is a glutamate agonist in the brain, meaning that it mimics your brain's primary excitatory neurotransmitter. Just like glutamate is decarboxylized into GABA (an inhibitory neurotransmitter) with the aid of the enzyme, glutamic acid decarboxylase, ibotenic acid decarboxylates into another bioactive compound, muscimol, which activates the brain's GABA receptors. The pure effect of muscimol is easy to study because that exclusively acts upon GABA receptors. However, the specific effect of ibotenic acid is difficult to parse from the effects of muscimol because any behavioral effects seen with ibotenic acid are indeed a mix of ibotenic acid and muscimol together.

lycoperdonosis (LY co pear don OH sis). Inhalation of large quantities of puffball spores can cause lung inflammation and respiratory distress. Lycoperdonosis has been observed in teenagers who were trying to 'huff' puffball spores (they ended up in the hospital) and in a few puppies (they later died).

muscarine (MUS car ine). Muscarine was originally isolated from *Amanita muscaria*, hence the name of the toxin. However, there is only a trace amount of muscarine in *A. muscaria*, and the psychotropic effects from this mushroom are primarily from muscimol and ibotenic acid. Muscarine poisoning is responsible for a group of symptoms that go by the acronym SLUDGE, referring to sweating, lacrimation (producing tears), urination, diarrhea, gastrointestinal distress, and emesis (vomiting). Essentially, fluids exuding from every orifice of the body, sometimes simultaneously, and severe reactions can result in death. Specifically, the toxin is an agonist to peripheral muscarinic acetylcholine receptors and leads to over-activation of the neurotransmitter. The antidote for this category of mushroom poisonings is atropine, an antagonist to these receptors. Specific mushrooms that can result in muscarine poisoning includes many *Inocybe* and *Clitocybe* species, including *Clitocybe rivulosa* (=*C. dealbata*), which has the common names 'sweating mushroom' and 'fool's funnel'.

muscimol (MUS kih mall): One of the two active chemicals found in *Amanita muscaria* and *A. pantherina* (as well as other species). Muscimol is a GABA agonist in the brain, meaning that it mimics one of your brain's two main inhibitory neurotransmitters. As such, activation is responsible for the tiredness, hypnotic, and dissociative effects of these muscimol-containing mushrooms. Muscimol is also responsible for mydriasis (pupillary dilation), which can be seen in the later stages of muscimol toxicity. Interestingly, the hallucinatory effects of this mushroom is mediated through the GABA receptor, and is not from the activation of serotonin receptors that underlie the classic hallucinogens, including *Psilocybe* species. See the entry for 'ibotenic acid' to see the relationship with muscimol.

orellanine (or EL lan ene). Orellanine was originally described from *Cortinarius orellanus*. Although the species was once considered edible in Europe, an outbreak in Poland in the 1950's led to over a hundred poisonings. Of these, there were 24 cases of kidney failure and 11 deaths. Across numerous studies, the rate of kidney failure is about 35%, and fortunately, about 60% of these people recover. Among all reports, the mortality rate after poisoning ranges between 5 to 15%, although death doesn't generally occur once dialysis has started, or if the patient has received a kidney transplant. Symptoms begin 2 to 30 days after consumption, and include nausea, vomiting, headache, and diarrhea before renal failure sets in. Unfortunately, this relatively lengthy latent period may cause the victim to be unsure of the cause of the symptoms. The most famous case of poisoning was the author of 'The Horse Whisperer', Nicholas Evans. He mistook these Cortinarius species for *Boletus edulis* and prepared them for dinner. He, his wife, and two of their friends all ended up getting kidney transplants. The toxin is in other related *Cortinarius* species, including *C. rubellus*, *speciosissimus*, *gentilis*, *smithiana*, *splendens*, and *armillatus*.

psilocybin: Psilocybin, Psilocin, and related chemicals are found in a number of species, including *Psilocybe cubensis* and *Gymnopilus junonius*. These mushrooms are hallucinogenic, and the active chemicals bind to and activate a specific type of serotonin receptor in the brain. R Gordon Wasson and Roger Heim were interested in the culture associated with hallucinogenic mushrooms and took several trips to Mexico (and trips *in* Mexico?). They were invited to participate in spiritual ceremonies where they consumed local psilocybin-containing mushrooms. The woman who led the ceremonies was Maria Sabina. The audio recordings of one of her ceremonies appears on the album 'Ceremony of the Mazatec Indians of Mexico'. Pictures of them all, including a popular press write-up of the mushrooms, was the focus of a famous LIFE story published in 1957. The phrase 'Magic Mushrooms' was coined when an editor included it as part of the article. Science is still exploring the use, benefits, and mechanism of action of the mushrooms. Kometer et al. (2013) gave a group of human volunteers either psilocybin or psilocybin with a 5-HT2A receptor antagonist, which functioned to block specific serotonin receptors. They found that the addition of this serotonin blocker resulted in those participants not experiencing the hallucinatory effects of the psilocybin (Kometer at al., 2013, Journal of Neuroscience volume 33, issue 25, pages 10544-51). It is activation of these serotonin receptors that underlies benefits seen in the common antidepressants, as well as the potential benefits from microdosing on psilocybin-containing mushrooms. Further, psilocybin-induced hallucinations arise from an over-activation of these receptors, just like too high a dose of antidepressants can also cause hallucinations themselves (termed 'Serotonin Syndrome').

rhabdomyolysis (rab doe my oh LIE sis): Rhabdomyolysis describes the breakdown of muscles, including muscle spasms, muscle pain, stiffness in the legs, and dark urine. Nausea without vomiting and profuse sweating are also symptoms. It can start 1 to 6 days after consuming large quantities of *Tricholoma equestre*. Across 12 reviewed cases, there were three deaths, and the others took 1 to 2 months to recovery.

silibinin: Silibinin is a chemical from milk thistle, *Silybum*, which is used in cases of amatoxin poisonings. Amatoxin-containing species like *Amanita bisporigera*, *Galerina marginata*, and *Lepiota felina* shut down the liver. Silibinin is an approved treatment for amatoxin-poisoning in Europe, but it is in Phase II clinical trials in the United States (at least at the time of this writing).

Yunnan Sudden Death Syndrome: The Chinese government ordered an investigation into 400 unexplained deaths in the Yunnan province over a 30-year span. The paper came out in the journal Science in 2010 identifying the culprit: a little white mushroom. After eating the mushroom, people would feel nauseous, have heart palpitations, and then suddenly drop dead from cardiac arrest. The species was *Trogia venenata*, now considered the world's deadliest mushroom (the epithet even means poisonous). More recent evidence points to two newly characterized amino acids in the mushroom that cause hypoglycemia and death in lab mice. Although the specific mechanism of action is unknown, a major educational campaign warning people not to eat the little white mushroom has apparently worked: there have been no additional deaths since the Science paper came out in 2010. Yay for science.

5. GENERAL GLOSSARY

Key to Understanding Entries

The following is a sample entry plus a key to understand it:

brunne (G): Brown. *Gymnopus brunneolus* (brun ney OL us) is brown, and *Gymnopus brunnescens* turns brown. *Amanita brunnescens*, which is probably better known than the *Gymnopus*, has brownish hues too.

word element (language of origin*): English translation. An example of a specific species that uses the word element (pronunciation). Typically, a description of how the word element is applied in the species is given, with other examples if appropriate. When possible, history or a few related words are also given.

* Latin is indicated by (L), Greek with (G), and English with (E). All other origins (French, German, Japanese, Russian, etc…) are spelled out in full.

A

a/an (G): Without. Used in *Pisolithus arrhizus* (ar RYE zus), to indicate the fungus is 'without roots'. This is a common prefix that is frequently used in science and in common language. For instance, 'amoral' means 'without morals', 'alexia' means 'without reading', and 'anencephaly' means 'without a brain'.

abies (L): Fir tree. *Trichaptum abietinum* (ah bi eh TIE num) grows on fir trees.

abort (L): To miscarry. *Entoloma abortivum* (uh BOR tih vum) was originally believed to be an improper (aborted) *Entoloma*. However, research presented in Czederpiltz, Volk, and Birdsall (2001) contradicts the original hypothesis.

abrupt (L): Abrupt. The epithet *abruptibulbus* (uh brup tih BUL bus) indicates that a stem has come to an abrupt end.

abscondit (L): Secret, hidden. *Pleurotus abscondens* (abs CON dens), and over 100 other species with a related epithet, grow out of plain sight. Related is 'abscond', which means 'to run away and hide'. To use them in a sentence: "Who absconded with my *Ramularia abscondita*!?"

abund (L): Abundant. You typically find *Clitocybula abundans* (uh BUN dens) in great numbers.

ac / acic (G): A point. Includes forms such as acicula and aculeata. *Collybia acicularis* (ay cic you LARE is), *Clavaria aculeate*, and *Mycena acicula*, are all pointed in some form or another. 'Acicular' is an adjective often used to describe pointy crystals, leaves, and needles. Common related words include 'acute' and 'acumen'. Also, *Accipiter* is a genus of hawks, and it literally means 'sharp wing'.

acanth (G): Spine, thorn. Used in nearly 200 records, including *Grifola acanthoides* (ay can THOY dees). Also, *Acanthodii* represents the spiny sharks, and *Acanthamoeba* is a spiny amoeba that can cause eye infections.

acer (L): Maple tree, the genus name of the maple tree. *Ciboria acericola* (ay ser ih COE luh) and *Stereum acerinum* are associated with maple trees.

acerbum (L): Bitter. *Tricholoma acerbum* (ah CER bum) is nicknamed the 'Bitter Knight'.

acerv (L): A heap. *Connopus acervatus* (ass air VAH tus) (formerly *Gymnopus acervatus*) grows in a clump (heap).

acetabulum (G/L): A vinegar cup used by ancient Greeks and Romans. *Helvella acetabulum* (ass eh TAB you lum) looks like one of the cups. The 'acetabulum' is also the point where the femur joins the pelvis, which is cup-shaped.

aculeatus (G): Having spines. *Gymnopilus aculeatus* (ay cue lee AH tus) and *Oudemansiella aculeata* both have spiny features. There are over 225 specific epithets that use this word element.

acumin (L): Pointed. Over 150 species use this epithet, including *Bovista acuminata*, which comes to a point. 'Acuminate' is an adjective to describe something that comes to a point.

adip (L): Fat. *Pholiota adiposa* (ad ih POE suh) has a fat pileus. In anatomy, 'adipose tissue' is body fat.

admirabilis (L): Worthy of admiration, wonderful. *Russula admirabilis* (add mir AB ill is) is just a wonderful mushroom. Despite my opinion that nearly every fungal species is wonderful, only nine have the epithet *admirabilis*.

adonis (G): Beautiful, the male Greek god of beauty and the archetype of handsome youth. *Mycena* and *Camarophyllus adonis* (eh DON is) are beautiful.

adstringere (L): To pull tight, also a dry puckering mouthfeel. Used in *Russula astringens* (as TRIN gens). Related to taste, astringency is the dry puckering mouthfeel from certain foods. *Russula astringens* has an astringent taste.

adusta (L): Burnt. *Russula adusta* (a DUS tah) has a burnt color to the cap.

aeger (G): Black poplar. *Cyclocybe aegerita* (ay GER ih ta) is a cultivated mushroom that grows on poplar logs, and it is commonly called the 'Black poplar mushroom'.

Aeminiaceae: After the Latin name of the city Coimbra, Aeminium. A new taxonomic family of fungi was described in 2019 growing on limestone artwork in an 800-year old cathedral in Coimbra, Portugal. Newly named was the family Aeminiaceae, genus *Aeminium*, and species *Aeminium ledgeri*. A note that the epithet was named after a late colleague of those who discovered the new fungus, Ludgero Avelar.

aeolus (G): Variegated, to vary. The genus name *Panaeolus* (pan ay OH lus) literally means 'all mottled', a word that highlights their mottled gills.

aeruginis (L): Copper rust, verdigris, blue/green. *Chlorociboria aeruginascens* (ay roo gin AYE sens) displays a green fruitbody. Although you are unlikely to find this fungus, you are very likely to find blue-stained pieces of wood on the forest floor if you look closely. This indicates the mycelium is in the wood and that the fungus will fruit under the right conditions.

aescul (L): An oak. *Trametes aesculi* grows on oak trees.

aestiv (L): Summer. *Boletus aestivalis* (es tiv AL is) (now *B. reticulatus*) and nearly 100 other species with similar epithets are associated with the summer.

affinis (L): Related or adjacent to. *Russula laracinoaffinis* (lar ah SEE no af FIN is) grows near Larch trees. In related terminology, the word 'affinity' means 'a natural liking' or 'relationship other than by blood'.

agath (G): Pleasant. *Hygrophorus agathosmus* (ah gath OHS mus) has one of the most pleasant smelling odors of any mushroom I have smelled.

agglutinare (L): Glueing. *Mycena agglutinatispora* has spores that clump together. Also, the crust fungus *Hymenochaete agglutinans* (ag GLU tin ans) (=*Hymenochaetopsis corrugata*) can grow across and fuse different sticks. In a more general sense, 'agglutination' is used in molecular biology to refer to cells that clump together.

aggregat (L): Assembled or heaped together, aggregated. The many species with the epithet *aggregata* grow in a clump.

agri (L): A field. *Ophiocordyceps agriotis* (ag ree OH tis) is named after the host, *Agriotes sericeus*, a wire worm that lives underground in fields. The word element is also used in the word 'agriculture'.

akanth (G): Spiny. The genus *Akanthomyces* covers moth species with spiny-white protruding fungal growths.

albida / albus/albulus (L): White. *Clathrus archeri var. alba* (AL ba), *Russula albidula* (al BID you luh), and *R. albida* are white. This root is related to the word 'albino'. Can you think of another mushroom name that utilizes *alba* in the species name?

alcalin (Arabic): Alkaline. *Gymnopus alkalivirens* turns green (virens) with an alkaline solution (you can use ammonia or KOH), while *Mycena alcalina* smells of ammonia.

Alessio, Carlo Luciano (1919 - 2006) was an Italian mycologist and author. He specialized in and published several works on Boletus and Inocybes. He described *Xerocomus ichnusanus* and *X. roseoalbidus*, five Inocybes, *Rubroboletus puchrotinctus*, and a few others. *Alessioporus* was erected in 2014 to elevate the *X. ichnusanus* Alessio had described to its own genus.

aleur (G): Mealy. *Aleuria* (ay LURE ee uh) *aurantia* (the orange peel fungus), has a mealy texture on the outer surface.

allant (G): Sausage. *Gymnopilus allantopus* (al LAN toe pus) literally means 'sausage foot': Berkeley must have been hungry when he named the species. Also, *Peniophorella allantospora* (al lan toe SPORE uh) has sausage-shaped spores.

Allen, Alissa: Alissa discovered *Pouzarella alissae* (uh LISS ay) on a November day in 2013 in California. Noah Siegel recalls her as saying "I found this thing that looks sorta like an Inocybe, but has pink spores". Noah knew it was a *Pouzarella*, but it was indeed a new species.

Allen, John W (1940 – current): The species *Psilocybe allenii* (al LEN ee eye) is named after John Allen because he "deeply believed in this new species and repeatedly insisted on a detailed study and DNA sequencing" (Borovicka et al. 2012, p. 185).

alligat (L): To bind to. Many species of fungi have epithets *alligata* (all ih GAT uh) and *alligatus*, like *Polyporus alligatus*, which binds to its substrate. Although the alligator binds to things it eats, it has a different etymology: it just means 'lizard' ('largato' is Spanish for lizard).

allium (L): Garlic, the genus name of garlic. *Marasmius alliaceus* (al lee AYE see us) smells like garlic.

alnus (L): Alder. Oddly enough, *Lactarius alnicola* (al NIC oh luh) is more common with Oak and conifer trees than alder.

altus (L): High. *Cantharellus altipes* (AL tih peas) has a rather long stipe and was described in 2011 growing with oak and pine in Texas.

aluta / alutarius (L): Leather, made of leather. *Russula alutaceiformis* (ay loo tay cee ih FOR mis), *R. alutacea*, and other similar-ending mushrooms all have a leather-like quality.

alveolus (L): A pit or small hollow. *Neofavolus alveolaris* (al veh oh LAR is), the 'hexagonal-pored polypore', has many small hexagonal-shaped pits.

amanita (G): An ancient term for a fungus, believed to be used first by Galen (129 AD to 200 or 216 AD). McIlvaine (1900) states that the term was derived "Perhaps from Mount Amanus", which is in northern Turkey. The pronunciation of *Amanita* varies widely across the United States. The most common pronunciation is (am an EE tuh), although (am an EYE tuh) is also used, particularly in the Rocky Mountains.

amara (G): Trench. *Russula amara* (uh MAR uh) has a trench around the center portion of the pileus. There are 130 hits on Indexfungorum.org for *amara* (both the Greek and Latin meanings), so this is a pretty common epithet.

amara (L): Bitter. *Lepista amara* (uh MAR uh) (formerly *Clitocybe amara*) tastes bitter.

amaur (G): Dark. *Amauroderma* (ay maur oh DER mua) and *Psathyrella amaura* are characterized by dark flesh.

amblys (G): Blunt, dull. *Inocybe amblyospora* (am blee oh SPORE uh) has blunt spores. In related words, amblyopia (lazy eye) results in poor (dull) vision.

americanum: Named after the United States of America, which was named after the explorer Amerigo Vespucci. Used in *Hericium americanum*, a prized edible.

amiant (G): Pure without spots. *Cystoderma amianthinum* (am ee an THI num) is unspotted. 'Amianthus' is a type of asbestos that is uniform and silky.

amicus (L): Friendly. Used in *Agaricus amicosus* (am ih COE sus). This mushroom was named in by Kerrigan (1989), who stated "The epithet *amicosus*, while suggestive of the gregarious habit of this mushroom, was actually chosen to celebrate the many friendships which are a hallmark of

Harry Thiers' career" (p. 124). Think of the word 'amicable', meaning 'friendly'.

amm (G): Sand. *Hygrocybe*, *Peziza*, *Psathyrella*, and *Conocybe ammophila* (am MOPH ill uh) all grow in sand. 'Psamm' is equivalent, and used in many other species.

amnicus (L): River. *Russula amnicola* (am NIC oh la) was originally described growing near a river.

amoenus (L): Pleasing, lovely. *Clavaria*, *Dermocybe*, *Peziza*, and *Russula amoena* (ay moe EN uh) are lovely mushrooms.

ampelinus (G): Of the grape vine. Used in *Russula xerampelina* (zer am peh LINE ah). Most sources state that *Russula xeramеplina* is named after the color of dried leaves of the grape vine. Vera Evanson (Mushrooms of Colorado) takes a different approach and notes the mushroom is named after the wine-colored cap.

ampl (L): Large, wide. *Tricholoma amplum* (AM plum) has a widening base. The epithet *amplissimus* is the superlative, meaning 'widest' or even 'most abundant'. Related words include 'amplify' and 'ample', among others.

ampulla (L): Bottle (ampula). Used in *Cordyceps ampullaceae* (am pull ASS ee ay), which is shaped like one of these bottles.

amygdal (G): Almond. Used in *Russula amygdaloides* (uh MIG dull OYE dees), which apparently looks like an almond. Also, *Hygrophorus amygdalinus* smells like almonds. In related vocabulary, the 'amygdala' is a nucleus in your brain that is shaped like an almond.

ana (L): Duck. *Russula anatina* (an uh TEE nuh) and *Entoloma anatinum* are colored like ducks (a brownish female duck). Anatidae is the Family name of ducks and *Anas* is the species name of the dabbling ducks.

ananas (L): Genus name of the pineapple. *Boletellus ananas* (uh nan us), the 'pineapple bolete', has pineapple-like scales on the cap.

ang (G): A vessel. The truffle *Hysterangium* (his ter AN gee um) appears to be a vessel for a womb (hyster-). Related words include 'angiosperm' (a casing for seeds), 'angina', and 'angiogram'.

angustus (L): Narrow/slender. *Russula angustispora* (an gus tih SPOR uh) has narrow spores.

annosum (L): Literally 'full of years'; perennial. *Heterobasidion annosum* (an NO sum) can be an old fungus. This is related to 'annual', but not 'anosmia' (without smell).

anomal (G): Irrregular. *Marasmius anomalus* has an anomalous shape.

anthrac (G): Coal, carbon. *Elaphomyces anthracinus* (AN thrah SEEN us) has a coal-black exterior.

antillar (L): Of the Antilles. Over a hundred species, including *Panaeolus antillarum* (an til LAR um), were originally described from the Antilles.

apal (L): Soft, tender. *Conocybe apala* (uh PAL uh) is rather delicate.

apio (G): A pear. *Apiosporina* (ay pee oh SPOR um) refers to pear-shaped spores and it is both a genus name and common specific epithet in the fungal world.

appendicul (L): To hang. Used in *Psathyrella appendiculata* (ap en dic you LAH tuh). Remnants of the partial veil are commonly left hanging off the margin of a mushroom's cap in many species. In a few species, it is the remnants of the universal veil that hang off the margin. Either way, it is a useful feature to describe when present. There are 248 records of mushrooms that contain the element *appendicul.*

applanatus (L): Flattened. *Ganoderma applanatum* (AP plan AY tum) is generally flat. Related words include 'plane' and 'planar'.

apric (L): Sunny, apricot-colored. *Amanita aprica* (AP rih cuh) was named for both its growth in direct sunlight and its apricot-colored pileus.

arachn (G): A spider. *Ophiocordyceps arachneicola* (ah RACH neh IC oh la) loves spiders, but people with arachnophobia don't.

arae / araios (G): Narrow. *Ramaria araiospora* has narrow spores.

arcan (L): Hidden. *Trametes* and *Psilocybe arcana* (ar CAN uh), and members of the genus *Arcangeliella* often appear hidden. 'Arcane' is a related word.

archae (G): Ancient. A prefix used in a few genus names to denote their age. For example, *Archaeomarasmius* was discovered in amber that was 90-million years old. Another is the Russula section Archaeinae. Related words are 'archeology' and the kingdom Archezoa.

Archer, William (1820 - 1874): *Clathrus archeri* (ARCH er eye) is the 'octopus stinkhorn', and many people would believe the 'archeri' refers to the mushroom's arching rays. In actuality, it is named after a person (it does end with an 'i', after all). This species was named after William Archer by Berkeley in 1859 (Dawson, 2013).

arcul (L): 'Arculus' was the Roman god of boxes and chests, while 'arcularius' refers to the profession of making those boxes. In botany and mycology, it is generally a reference to being box-like. *Polyporus arcularius* (ar cue LAR ee us) was named after its box-like pores.

arcy (G): A net. *Arcyria* (ar SEER ee uh), the genus that includes the 'cotton candy slime mold', is named after the network of fibers connecting the sporangia.

aren (L): Sand. *Scleroderma arenicola* (are en ICK oh luh) grows with sand.

areolate (L): Structure with a network of cracks, like dried mud. *Scleroderma areolatum* (ar eh oh LAY tum) develops fine cracks with maturity. *Lactarius areolatus* has spores that appear to be finely cracked.

argent (L): Silver. Used in the epithet *argenteus* (ar GEN tee us) and *argentea* to denote color. These epithets are used with *Agaricus*, *Coprinus*, and *Phlebia*, to name a few. In related terminology, the country 'Argentina' is derived from the same meaning, and the abbreviation of silver on the periodic table of the elements is an abbreviation of 'argentum', 'AG'.

argil (G): White clay (reference to color). Used in *Bovista* and *Clavaria argillacea* (are gill AH see aye), as well as over a hundred more species.

argyros (G): Silver. *Russula argyracea* (ar gye rah see uh) is silvery in color.

aria/arius/arum/arium (L): Relating, pertaining, or belonging to. This is a common suffix denoting an association with the preceding portion of the word. This may include an association with another species, but can also broadly draw emphasis to a key feature. The suffix is used in the genera *Lactarius* (pertaining to milk) and *Cortinarius* (having to do with a cortina), and far too many specific epithets to consider.

armeniac (L): Apricot, apricot-yellow. In addition to being the genus name of the apricot, *armeniaca* is used in over 100 species of fungi, including *Amanita armeniaca* (ar men ee AH cuh), the 'gypsy peach amanita', which should probably be called the 'gypsy apricot amanita'. 'Armenia' is the origin of the apricot.

armilla (L): Ring, bracelet. *Gymnopilus armillatus* (ar mil LAT us) and many *Armillaria* species have a ring on the stem, except the ringless ones of course.

aroma (G): A spice, fragrant. *Gymnopilus aromaticus* (ah rho MAT ick us) is fragrant.

Arora, David (1952 - current) is an American mycologist and author. Arora has published two books, "Mushrooms Demystified" and "All That the Rain Promises and More", which are arguably two of the best-selling mushroom books out there. He also founded the Fungus Federation of

Santa Cruz, circumscribed *Butyriboletus*, and named species including *Cantharellus californicus* and *Craterellus atrocinereus*. *Agaricus arorae* was named after him by Rick Kerrigan, who nostalgically credits Arora for being the first person to introduce him to mycology. The word on the trail is that a second edition of "All That the Rain Promises and More" is coming, followed by a third edition of "Mushrooms Demystified". I think an memoir would be a hit, too.

Arrhenius, Johan Peter (1811 - 1889): Arrhenius was a Swedish botanist, and the genus *Arrhenia* (uh RAIN ee uh) was named after him.

arven (L): In the field. Used in *Agaricus arvensis* (are VEN sis), which we call 'the horse mushroom' for somewhat unclear reasons. *Agaricus campestris*, also translates to 'the field mushroom', must have claimed the common name first.

arvern (L): After Auvergne, France. *Peziza*, *Tricholoma*, and *Puccinia arvernensis* were all described from this region in central France. Confusingly, *Peziza arvernensis* is also found in North America.

aseroe (G): Disgusting juice. Members of the genus *Aseroe* (ASS er oh or ASS er ooh) are commonly called the 'starfish fungi' or 'anemone stinkhorns'. These stinkhorn species have a smelly and sticky spore surface.

asper (L): Rough. *Echinoderma asperum* has rough scaly caps.

aspergill (L): A holy water sprinkler. Conidia of *Aspergillus* (as per GIL lus) resemble an aspergillum, something priests use to sprinkle holy water. It isn't the first thing that pops into my mind when I see the conidia, but it was an Italian priest with a penchant for botany who first described and named the genus.

aspid (G): A round shield. *Lactarius aspideus* (as PID ee us) and *L. aspideoides* are reminiscent of an 'aspis', the round wooded shield of ancient Greece.

aster (G): Star. Astraeus (as TRAY us) are broadly referred to as earthstars after their star-like rays.

aster (L): A suffix added to a noun to indicate a diminutive form. For example, the genus *Flammulaster* (FLAM you last er) looks like a little flame.

ater/atr (L): Black. *Phallus atrovolvatus* (at rho vole VAH tus) literally means 'a penis with a black covering at the base'. There are a several other species names that start with *atro* that indicate something about them is black.

athena (G): Athena, the daughter of Zeus and protector of Athens. The genus *Atheniella* (ath en ee EL uh) was named after Athena, both because of the beauty of the fruitbody and their fancied resemblance to a shield and spear. Note that *Mycena adonis* was renamed *Atheniella adonis.*

atomar (L): Covered with spots. Species like *Psathyrella atomatoides* and the lichen *Leptorhaphis atromaria* (ah toe MAR ee uh) appear to be covered in spots.

atractos (G): Spindle-shaped. *Atractosporocybe* (ah trac to spor AH cih bee) is a mouthful, but it just means the mushroom has spindle-shaped spores and is generally Clitocybe-like. *Atractospora* is also an epithet used in a couple of fungal species to describe the shape of spores. *Atractosteus* is a genus of gars, and perhaps the best-known example is *A. spatula*, the alligator gar.

atramentum (L): Ink, black. *Russula atramentosa* (ah trah men TOE suh) and *Coprinopsis atramentaria* are both blackish.

august (L): Notable, majestic. *Amanita augusta* (au GUS tuh) is known as 'the majestic amanita'. The month of August was named after Augustus Caesar, who was given the name Augustus by the Roman senate.

aurant (L): Orange. *Hygrophoropsis aurantiaca* (are an TIE uh cah) is a common orange mushroom, and is called the 'false chanterelle'. There are several species of false chanterelles, though, which is one reason I am not a fan of common names. This *Hygrophoropsis* has gills instead of the gill-like ridges of the chanterelle, but people still apparently confuse them.

auricula (L): External part of ear. Used in *Auricularia* (au ric you LARE ee ah). Auriculariaceae is a taxonomic family of jelly fungi. *Auricularia* is a species of jelly fungi, most notably *Auricularia auricula-judae*, the wood-ear jelly. In related terminology, the 'auricle' is the external part of the ear.

aurum (L): Golden, gold colored. *Cantharellus aurora* (uh ROAR uh) is commonly called the 'golden chanterelle' (and has nothing to do with mycologist David Arora, whose last name is spelled slightly different).

austr (L): Of the south wind. *Austroboletus* (au stro bo LEE tus) literally means the 'Southern Bolete', but they can occur as far north as Canada (at least southern Canada anyway). 'Australia' is the southern continent.

avellaneus (L): The color of a fresh filbert or hazel-nut shell. Used in *Russula cremeoavellanea* (crem EH oh av ell an EE ah). If this had a common name, it would be the 'Filbert Russula'. The Hazel tree is *Corylus avellana.*

ax (G): Cylinder, axle, axis. The genus *Podaxis* (*Podaxon* in older literature) essentially refers to the stalk a cylinder. In other terminology, the 'axon', the projecting arm of a neuron, is also cylinder-shaped.

azur (L): Azure, sky blue. The stem of *Psilocybe azurescens* (az or ES cens) stains blue with handling. This mushroom contains psilocybin and is hallucinogenic.

B

Badham, Charles David (1805 - 1857) was an English physician whose interest in mycology started with mycophagy. In 1847, he published a book, 'A treatise on the esculent funguses of England'. He sent many collections to Berkeley, and as such, had some species named after him. The slime mold genus *Badhamia* (bad HAM ee uh) was named after him, and among the fungi, *Leucocoprinus badhamii* was named after him.

badius (L): Chestnut-colored, brown. *Phellinus badius* (BAD ee us) and *Polyporus badius* are both brown.

bae / bai (G): Small, slim. Used in *Arrhenia baeospora* (bay oh SPORE uh) and *Leucopaxillus baeospermus* to describe the spores. *Baeomyces* is also a genus of lichen.

baios (G): Small, slender. *Arrhenia baeospora* (bay oh SPOR uh) has small spores. This word element is used in nearly 100 specific epithets.

balan (G): An acorn, acorn-shaped, equivalent to the Latin 'glans'. There are over 150 species that start out with 'balan'. These species look like an acorn or just grow under oak trees, like *Mycena balanina* (bal an EE nuh). The 'balan' is another name for the glans of the male reproductive organ.

balen (L): Whale. *Myrtapenidiella balenae* has no resemblance in any way to any feature of whales. If there was an award given to the most immaterial epithet ever given, this would win. This species was named after the fact that whales were "present close to the shoreline at the time that this fungus was collected". Collecting species near an ocean with a warm breeze and whales breaching nearby sounds nice, but can we all just agree that binomials should in some way relate to the actual species discovered?

Ballou, W. H. (1857 - 1937) was an American mycologist who frequently sent specimens to Peck and Lloyd. Ballou made the original collection of what we now call *Tylopilus balloui*, and a dozen or so other species are named after him as well. It should be noted that the epithet *balloui* is corrected from *ballouii*.

balteat (L): Girdled, belted, banded. *Cortinarius balteatus* (ball tay AH tus) has an annular zone near the apex of the stalk that has picks up a heavy spore deposit.

Banker, Howard James (1866 - 1940): Banker was an American Mycologist who specialized in the Hydnaceae. He was a founding editor of Mycologia in 1909, and published 8 papers on the family in the the journal between 1912 and 1929. The family Bankeraceae and genus *Bankera* (erected in 1955) were named after him.

Banning, Mary Elizabeth (1832-1903): was one a few early women mycologists. She originally discovered species like *Amanita banningiana* (provisional name), *Hypomyces banningiae*, and *Rususla mariae* (Mary's Russula). She also published a number of works, and did some amazing watercolors. She did 174 watercolors for a book 'Fungi of Maryland' that was never published.

bao (Chinese): Thin. The genus *Baorangia* was circumscribed in 2015 to include some boletes, and it is named after the thin pore surface. Etymologically, it is bao + rang + ia. Thus, an acceptable pronunciation would be bow RANG ee uh).

baph (G): Dyed. *Cortinarius dibaphus* (die BAPH us) is twice dyed, while *Amanita hemibapha* is half-dyed.

barbatus (L): Bearded. *Coltricia barbata* (bar BAH tuh) has a bearded margin. Related words include barber and barbs. The country Barbados got its name because of the bearded fig tree. Barbate can also be used as an adjective to describe something that is bearded, like 'my face is a little barbate today'.

Barrows, Charles (Chuck) (1904-1989): Chuck was an amateur mycologist from the southwest United States. Smith and Thiers named *Boletus barrowsii* (bar ROW zee eye), the so-called white porcini that is common in states like New Mexico and Colorado, after him in 1976.

bas (L): Base. The base of the stem of *Russula luteobasis* (loo tee oh BASE is) is yellow. In other terminology, you can think of the word *basal*, meaning *bottom*, including the *basal ganglia*, a brain structure near the base of the brain.

bathy (G): Deep. *Poria bathypora* (bath ee POR uh or bat he POR uh) has deep pores, and *Mycena bathyrrhiza* has deep roots.

batr (G): Frog. *Batrachochytrium* (bah tray cho CHY tree um) *salamandrivorans* is a fungus whose zoospores develop in flower-pot shaped zoosporangia (chytrium means flower-pot shaped). It results in an infectious disease in amphibians, and was originally described in frogs, although *Batrachochytrium salamandrivorans* affects salamanders. A 'batrachian' is an amphibian.

Battarra, Giovanni Antonia (1714 – 1789) was an Italian priest, naturalist, mycologist, and author. The genus *Battarrea* and the species *Psathyrella battarrae* were named after him.

Beauverie, Jean (1874 – 1938) was a French mycologist. In 1912, *Beauveria*, a genus of entomopathogenic fungi was named after him. The type species, and perhaps the best known example, is *Beauveria bassiana.*

bellus (L): Charming, pretty. Used in *Gymnopilus bellulus* (bell LOO lus). With the diminutive suffix *–ulus*, *Gymnopilus bellulus* is a pretty little mushroom. In related terminology, you may be familiar with the Italian expression *ciao bella*, meaning *hi beautiful.*

Berkeley, Cecilia Emma (1810 - 1881): Wife of Miles Joseph Berkeley, honored forever with species *Amanita ceciliae* (ceh SEAL ee ay). It was named after her in 1854 (as *Agaricus ceciliae*), and Berkeley stated "The name is intended to record the services which have been rendered to Mycology by many excellent illustrations and in other ways by Cecilia E. Berkeley".

Berkeley, Miles J (1803-1889): There are a mind-blowing 6,000 species of Fungi credited to him, including *Bondarzewia berkeleyi* (Berkeley's polypore). This is probably one of the top 10 species that comes up most often on the Facebook identification groups.

betula (L): Birch. *Piptoporus betulina* (bet you LIE nuh) is a common polypore that grows on birch trees. The genus of the birch tree is *Betula* and the family is Betulaceae.

bi (G): Two. *Amanita bisporigera* (bi spore ih GER uh) has two spores on each basidium.

Birnbaum: Birnbaum was a garden inspector in 1839, and discovered a mushroom in a greenhouse that now bears him name. *Leucocoprinus birnbaumii* (burn BAU me eye) is among the most frequently posted mushrooms on the identification boards.

Blakesley: After Albert Francis Blakesley. Most species are named after someone to honor them. Not *Phycomyces blakesleeanus* – Burgeff created this name as an insult to Blakesley. Fortunately, the epithet *blakesleeana* was also used and appears in several species.

blandus (L): Smooth. *Russula blanda* (BLAND uh) has smooth spores, but I cannot confirm that this is the particular feature denoted by the epithet.

blatta (L): A genus of cockroaches. *Ophiocordyceps blattae* (BLAH tae) grows from adult cockroaches.

blewit (E): The blewit, *Clitocybe nuda*, is bluish-purple when young. The old English common name was 'blue hat', but you will have to say this phrase 10 times fast and pretend you're British to come up with 'blewit'. Dictionary.com says it came into usage between 1820 and 1830, but the earliest reference I found was from the 1847, and even then it was mentioned as it was a common nickname.

Bloxam, Andrew (1801 - 1878) was an English naturalist who joined a two-year expedition on the HMS Blonde when he was just 22. Notably, he made a collection of a bird species in Hawaii in 1825 that later went extinct. His collection, which has survived, is the only existing collection of the species. As for fungi, he became associated with Berkeley, who named *Agaricus bloxamii* after him in 1854 (it was renamed *Entoloma bloxamii* in 1887). His name is in a number of other epithets as well, and the genus of coelomycete *Bloxamia* is named after him. Finally, he authored *Peziza berkeleyi*.

Blumenau: A city in Brazil, named after a German who settled it in 1850. The genus *Blumenavia* (blue men AY vee uh) was named by Möller in 1895 to pay homage to the city where he lived for a number of years studying phalloid fungi. Thus, the common stinkhorn *Blumenavia angolensis* is named after two places, Angola where it was actually found, and Blumenau where it wasn't found.

bolaris (G): A lump or clod, lumps of paint. *Cortinarius bolaris* (bo LAR is) appears to be speckled with red paint.

bolbit (G): Cow dung. Members of the genus *Bolbitius* (bol BIT ee us) grow on well-decomposed dung or in very fertilized areas.

bolet/bolus (G): Lump (a lump of clay), clod (a lump of earth), in the general sense, a mushroom. Sometimes taken to mean a superior kind of mushroom. Used in the genus *Boletus* (boe LEE tus), and in *Xylobolus* (zie low BOL us) *frustulatus*, the ceramic parchment, which appears as many individual clumps of ceramic on decomposing hardwoods.

bombyc (L): Silkworm. *Volvariella bombycina* (bom buh SEE nuh) has silky fibers on the cap that are most noticeable on the margin. It is sometimes called the 'silky rosegill' mushroom.

Boone, William Judson (1860-1936): Boone was a theologian, botanist, and the first president of the College of Idaho. He was a minister and officiated the wedding of Alexander and Helen Smith. Boone introduced

Smith to *Calvatia booniana* (Western giant puffball), and Smith named it after him (Kuo, 2008).

borealis (L): Northern. Used in *Climacocystis borealis* (bor ee AL is). The word *boreal* can be used to describe a species found in more northern regions. Thus, northern-occurring Russulas or other species can be described as being boreal. *Borealis* is the epithet in several genera, including *Armillaria*, *Climacocystis*, and I*nocybe*, to name a few of the common ones.

Both, Ernst E (1930 - 2012) was an expert on boletes and a president of the Buffalo Museum of Science. He published "The Boletes of North America: a compendium" in 1993, and in 2007, the genus *Bothia* (both ee uh) was named after him. *Bothia* is a monotypic genus, containing only *Bothia castanella*, which is known from the east coast of the United States and across Pennsylvania to Ohio.

botry (G): A cluster, including a cluster of grapes. *Glomus botryoides* (bah tree OYE dees) has spores that are clustered like grapes. This word element is used in over 300 fungal species.

bovista (German): According to Merriam-Webster, Bovista comes from the Middle High German word *Vohenvist*. This is a combination of two words, *vohe*, meaning *fox*, and *vist*, meaning *emission of gas from the colon*. Essentially, *bovista* means *fox flatulence*. Bovista (Boe VIS tuh) is a genus of puffballs. The website first-nature.com states that *bovista* refers to the smell of the released spores. Did someone really know what the smell of a fox's fart smelled like? You should not be surprised a mushroom is named after a fart though. Most people (ok, maybe not most people, but most super nerdy myco-friends) know another organism named after a fart. The name of the slime mold *Lycogala epidendrum* means *wolf-fart*. This is apparently what late nights in the mycology lab will do to you. And now you know two organisms named after a fart. Back to *Bovista*: *Scleroderma bovista* and any number of species within the genus *Bovista* are examples of the fox-fart in use.

brachy (G): Short. *Geopyxis brachypus* (BRACK ee pus) has a short stem.

brassic (L): Cabbage. *Gymnopus brassicolens* (brass IC oh lens or bras sih CO lens) has the foul odor of rotting cabbage, although a few reports equate it to the smell of a dead animal. *Brassica* is the genus name of the cabbage.

Bresadola, Giacopo / Giacomo (1847 - 1929) was an extremely well published Italian mycologist. The genus *Bresadolia* is named after him, and there are over 200 species that use his last name in their epithet, like *Amanita bresadolae* and *Inocybe bresadolana*. The list of genera and species named by him is equally impressive. He originally described the genus

Copelandia, and about 30 total species. His publications contained beautifully accurate depictions of the fungi he described.

brevis (L): Short. *Russula brevipes* (BREV ih peas) has a short stalk, and *R. brevis* is just generally short. In related terminology, an orator who values brevity is going to keep the talk short.

bruma (L): Winter, the winter solstice. *Brumalis* (broo MAL is), pertaining to winter, is used as an epithet in *Volvariella*, *Lentinus*, and many other species that are found in the winter. In related terminology, 'brumalia' is an ancient Roman festival celebrating the winter solstice, and 'brumation' is a state of sluggishness that occurs in cold-blooded animals in very cold temperatures that is not as severe as hibernation.

brunne (G): Brown. *Gymnopus brunneolus* (brun ney OL us) is brown, and *Gymnopus brunnescens* turns brown. *Amanita brunnescens*, which is probably better known than the *Gymnopus*, has brownish hues too.

bryon (G): A lichen or tree-moss. *Gymnopilus bryophilus* (bry AH phil us) grows on decayed wood with moss. In other terminology, a bryophyte is a term used that encompasses mosses and liverworts.

Buchwald, Niels Fabritius (1898 - 1986): The genus *Buchwaldoboletus* includes about 15 species of the Boletaceae, and was named by Pilat in 1969. The genus was named after Buchwald, a Danish mycologist and author. In an ironic twist, the prefix 'buchwald' is also German for 'beech-wood', and this had led to a lot of confusion (among Germans, at least) because *Buchwaldoboletus* are typically associated with conifers. Indeed, the common North American species *B. lignicola* is a parasite on *Phaolus schweinitzii*, itself associated with conifers.

bufo (L): A toad. Used in *Tricholoma bufonium* (boo FOE nee um), whose pileus somehow resembles a toad's back.

bulg (L): A leather sac. Members of the genus *Bulgaria* look like small leather sacs. The name of the country Bulgaria is from an unrelated Turkish word.

burs (L): Bag, purse. Epithets like *bursa* (BUR suh), *bursum*, and *bursiformis* refer to a bag-like fruitbody or structure. Related, the 'bursar' at a college deals with bags of money, or at least they used to.

butyr (G): Butter, buttery. *Clitocybe* and *Rhodocollybia butyracea* (byou tir RAH see ay) are greasy like butter. In related terminology, butyric acid is found in range of foods, including butter, cheese, milk, and chocolate. Interestingly, the characteristic smell of vomit also comes from butyric acid.

byssos (G): A fine yellowish flax. Members of the genus *Byssonectria* (biss so NEC tree uh) are yellow.

C

caeruleus (L): Dark-blue. *Craterellus caeruleofuscus* (see rue lee o FUS cus) has bluish shades. It is found in sphagnum bogs in the Great Lakes region, growing scattered to gregarious, but not in dense clusters.

caesi (L): The gray of the eye, a bluish gray. *Postia caesia* (SEE zee uh) is called the 'blue cheese polypore' and the epithet comes from the color of the fruitbody. The element 'Cesium' is also named after this color.

Cajander, Aimo Kaarlo (1879 - 1943) was a Finnish botanist. He was a professor of forestry at Helsinki University until he became a Director General for Finland's Forest Service. To top that, he was also the Prime Minister for stints in 1922 and 1924, and then again from 1937 to 1939. He was mostly interested in vascular plants, and also collected fungal parasites on those plants. This included what would become *Fomes cajanderi* (now *Fomitopsis* or *Rhodofomes cajanderi*), which was named after him in 1904.

calamistr (L): Curling, curling iron. *Inocybe calamistrata* (cal uh mis TRAH tuh) has a notably curled margin. *Calamus* (derived from the Calamus in mythology) can also mean 'reed' and is a genus of palms, and a genus of fish that live near reeds.

calic (G): Cup-shaped. *Phaeocalicium* (phay oh cal LIS ee um) are saprophytic lichens marked by dark gray (=Phaeo) cup-shaped apothecia. *Phaeocalicium polyporaeum* is found on *Trichaptum biforme*. A special thanks to Gerry McDonald from the Mid-Hudson Mycological Association for pointing this one out on a Sunday walk a few years ago. 'Calicular' is an adjective for something cup-shaped, and the 'chalice', a goblet, has the same origin.

caliga (L): A boot, a soldier's boot. Used in *Tricholoma caligatum* (cal ee GAT um), *Peziza caligata*, *Cortinarius caligatus*, and *Mycena caliginosa*, to name a few. *Tricholoma caligatum* is named "from a fancied resemblance in the stem to a leg with soldier's shoe" (British Basidiomycetes, WG Smith, 1908). A Roman caliga (plural = caligae) is not exactly the first thing I think of when I see this mushroom.

call / callim / calo (G): Beautiful. The epithet of *Hyphoderma cryptocallimon* (crip toe CAL lim on) literally means 'hidden beauty'.

callos (L): Callous, hard skinned. *Peziza* and *Poria callosa* have thick-skin. Related words include 'callous', the 'corpus callosum' of the brain, 'callus', and 'callousness'.

calvatus (L): Bald. Members of the genus *Calvatia* (cal VAY shuh) are nice and smooth.

calyc / kalyc (G): A calyx or cup. *Craterellus calyculus* (cal ICK you lus) is another redundant name, and it literally means, 'little cup little cup'.

calyp (G): Covered, hooded. Used in the epithets *calyptriformis* (cah lip trih FOR mis) and *calyptrate*, which are used in almost 70 species names. There is also a Greek nymph called 'Calypso' (hidden knowledge), but 'calypso' the types of music has a different etymology and is unrelated to this meaning.

camar (G): Vault. *Camarophyllus* (cam ar AH fill us) literally means 'vaulted gills'.

camillea (named after Camillea Montagne, the French Botanist): According to McIlvaine, Camillea Montagne proposed calling a new genus of Xylariaceae *Bacillaria*, but that name had just been claimed so Fries suggested he name it after himself. And so he did, and that is how the genus *Camillea* was born.

campanula (L): bell. *Panaeolus campanulatus* (cam pan you LAT us) has a bell-shaped cap.

camphor (Sanskrit): Camphor tree. Used in 60 species, including *Cortinarius* and *Lactarius camphoratus* (cam for AH tus), which have the odor of camphor. Medicinal VapoRub is 5% camphor, and moth balls are sometimes made of camphor.

campus (L): Field/plain. *Russula campestris* (cam PES tris), *Agaricus campestris*, and many others (there are 184 records of 'campestris' on Indexfungorum!) grow in fields or plains. The college 'campus' is related.

campyl (G): Bent. *Plectania campylosporus* (cam pill oh SPOR us) has bent spores, and *Crepidotus campylus* has crooked pileocystidia.

camur (L): Crooked. *Cortinarius camurus* (CAM ur us) (now *C. valgus*) has a crooked stem. Other species use *camura* and *camurandrum*. 'Camurus' is related to the Greek 'campyl', and has the same meaning.

cancellus (L): Lattice-work. *Clathrus cancellatus* (can cell AT us) is actually redundant, as *clathrus* also comes from a word meaning lattice-work too. Although this mushroom is now called *Clathrus ruber*, the name appears in older texts (including Arora, 1986) as *C. cancellatus*.

candid (L): Clear, white, shining. *Marasmiellus candidus* (can DID us) is a beautiful pure white. Related, a 'candid photograph' is pure and sincere.

de Candolle, Augustin Pyramus 1778- 1841 (Swiss): *Psathyrella candolleana* (can doe lay AN uh) was named after him. de Candolle had a notable career and posed many general scientific hypotheses that were later proven to be correct. He was the first to propose that we have an internal biological clock. He also promoted what we now call convergent evolution, that many species can evolve the same adaptations, which influenced Darwin and his theories on evolution. de Candolle also introduced the word 'taxonomy' in 1813, and made significant contributions to botany and mycology throughout his life. At least one source (British Basidiomycetes, WG Smith, 1908) states that the species was named after Augustin's son, Alphonse de Candolle.

cantharus (L): A deep cup of ancient Greece with a high stem and loop-shaped handles continuing the curve of the bottom of the body and rising above the brim (From Merriam-Webster.com). Used in *Cantharellus* (CAN thar EL us), the genus of chanterelles. When you find a chanterelles, think of the cup they were named after.

canus (L): White or gray. *Cantharellus roseocanus* (rho zay oh CANE us) has a light pinkish cap when young, and it turns pale with age and/or sunlight. It is a common choice edible in the Rocky Mountains.

caperat (L): Wrinkled. *Cortinarius caperatus* (cap ar AY tus) can have a wavy or wrinkled pileus. The etymology of the common name of this mushroom, 'the Gypsy mushroom', is unknown.

capill (L): Hair. *Conocybe* and *Mycena capillaripes* (cap ill AIR ih peas) have hair-like stems. The capillary tube got its name because a hair can fit through the center of it.

capitatus (L): Having a head. *Elaphocordyceps capitata* (cap ee TAH tah) is commonly called the 'drumstick trufflehead'. This one grows from an underground deer truffle. As the name suggests, it has a head on it.

capn (G): Smoke. *Hypholoma capnoides* (cap NOY dees) has smoky-grey colored gills.

capr (L): Goat. *Albatrellus pes-caprae* (pez CAP rae) is called the 'goat's foot Albatrellus'. Even though it is now *Scutiger pes-caprae*, 'goat's foot Scutiger' hasn't caught on yet.

capsicum (L): The genus name of a pepper, or spicy. Species like *Lactarius capsicum* (cap SIC um) are spicy, while *Cercospora capsici* is a fungal pathogen on the leaves of the pepper plant. 'Capsaicin' is the spicy component in peppers.

carbon (L): Charcoal (color). *Pholiota carbonaria* (car bun AIR ee uh), as well as many others with the same epithet, are carbon-colored.

carchar (G): Jagged. *Cystoderma carcharias* (car CHAR ee as) is known as the 'pale granular Lepiota' because it has jagged granules on the pileus. *Carcharias* is both a genus of tiger sharks (jagged/sharp teeth) and the specific epithet of the great white shark.

cardinalis (L): Chief, principal, red. *Cordyceps cardinalis* (car din AL is) was named its red stromata. This is a species described from eastern North America and East Asia by Sung and Spatafora (2004). The cardinal (the bird) is also derived from this word for red.

carios (L): Decayed. Species with the epithet *cariosa* (car ee OH suh) refer to a rotten-looking or smelling species. Related words include 'carrion' (decaying dead animals) and 'caries' (dental cavities).

carmine (French): A vivid crimson. It is used in a number of species including *Lactifluus carmineus* (car MIN ee us) *and Boletus carminiporus.* 'Carmine' is also used as an adjective.

carn (L): Flesh-colored, fleshy. *Ganoderma carnosum* (car NO sum) has a fleshy margin, and *Russula carnicolor* is generally flesh-colored.

carota (L): Carrot. Both *carota* and *carotae* are commonly used as epithets. *Inocybe caroticolor* is the color of a carrot. 'Beta-carotene' is also carrot-colored.

carp (G): Fruit. Used in sporocarp (SPORE oh carp), which is another word for a fruitbody. The word element is also used in a number of specific epithets too.

cartilaginis (L): Cartilage. *Exidia cartilaginea* (car til adge in EE uh) feels like cartilage.

caryophyllea (L/G): The specific epithet of the carnation, *Diathus caryophyllus*, itself named after the color of flesh. *Thelephora caryophyllea* is named after a general resemblance to the carnation, while other species with the epithet are specifically named after the color of the carnation.

castanos (G): The chestnut tree (Genus = *Castanea*), chestnut brown. The cap of *Russula castanopsidis* (cas tan OP sid is) is chestnut-brown. *Lepiota castanea* is another common mushroom utilizing this word element.

castellae (L): The Latinized form of 'Castella', an area in Shasta county, California. *Amanita castellae* (cas TELL ay) was originally found in Castella.

castor (L): A beaver. *Lentinellus castoreum* is the color of a beaver. Fries wanted this beaver Lentinus to fit with the bear (*Lentinus ursinus*) and fox (*Lentinus vulpinus*) Lentinus.

catinus (L): A deep serving dish, a large bowl. *Clitocbye* and *Tarzetta catinus* (ka TI nus) are shaped like wide, deep bowls.

cauda (L): Tail. *Hebeloma longicaudum* (lawn gee CAW dum) was named for its long stalk (long tail). A related phrase 'cauda equina', which is the name of a plant and a portion of our spinal cord. 'Caudal' is also a anatomical direction, and is a synonym of posterior.

cavus (L): Hollow. *Russula* and *Suillus cavipes* (CAV ih pees) have a hollow stem.

cedr (G): Cedar. *Russula cedriolens* (sed ree OH lens) smells like cedar.

cep (L): An onion. *Scleroderma cepa* looks like an onion, *Leucocoprinus cepistipes* means 'onion stem', and *Allium cepa* literally means 'garlic-onion'.

cep / cepe (Catalan / French): Trunk. Cep is one of the many common names for *Boletus edulis*, and this one refers to the thick stem of the mushroom.

cephal (G): Head, or as it pertains to mushrooms, the cap. *Gymnopus iocephalus* (eye oh ceph AL us) has a violet (io-) cap. There are a whopping 193 mushroom species that use the word element 'cephal' in their name! In a related word, 'hydrocephaly' is an unfortunate condition where the head becomes filled with cerebrospinal fluid (water).

cer (G): Honeycomb. *Cerioporus* (cer ee oh por us) *squamosus* (formerly called *Polyporus squamosus*) has dense honeycomb-like pores.

cercopes (G): A pair of mischievous forest creatures, who were difficult to pin down. Used in the genus *Cercopemyces* (ser co peh MY sees), a genus named in the summer of 2014 that proved elusive to mycologists. Not coincidentally, *Cercopemyces* was found under mountain mahogany, *Cercocarpu*s.

cere (L): Wax. *Peziza cerea* (CER ee uh) is yellowish and has a waxy appearance.

cerrena (Italian): A name for a fungus. Used in the genus *Cerrena* (ser REN uh), and a common example is *Cerrena unicolor.*

ceruss (L): White lead. Species like *Ceriporiopsis cerussata* (cer rus AT uh) and *Marasmius cerussatus* have the color of white lead.

cervinus (L): Deer, tawny like a deer. Many sources indicate that *Pluteus cervinus* (cer VINE us) is named for the color-similarity to deer. However, it appears that it was named after their antler-shaped cheilocystidia (sterile cells on the edges of the gills). Deer are in the taxonomic family Cervidae, subfamily Cervinae, and some are in the genus *Cervus.*

cessan (L): Delaying, late. *Russula cessans*, the 'Late Russula', is a late fall and early winter Russula in the eastern United States. 'Cease' is related.

chaet (G): Long flowing hair, a mane. The family name Hymenochaetaceae (high men oh chee TAH cee aye) means 'a hairy membrane'.

chaga (Russian): Derived from word for fungus. Chaga (CHAH ga) is the common name for *Inonotus obliquus*, and it just means 'fungus'.

chalc (G): Copper. The genus *Chalciporus* (chal cih POR us) has copper-colored pores.

chalyb (G): Steel. *Entoloma* and *Leccinum chalybeum* (kuh LIB ee um) have various shades of steel-gray, while *Entoloma chalybescens* turns that color. In words you probably don't come across often, 'chalybeate' water has iron salts, and 'chalybeate' refers to the color of steel.

chamaele (L): A chameleon, from the genus *Chamaeleo*. Species like *Lactifluus chamaeleotinus* and *Russula chamaeleon* change colors.

chelid (G): A swallow, which arrives when the yellow flowers of *Chelidonium* bloom, and leaves when they wilt. Later taken to refer to the yellow color itself. *Lactarius chelidonium* (chel ih DOE nee um) has distinct shades of yellow.

chicago (Algonquian): Wild onion, which are found in great numbers in and around Chicago. Used in *Canatharellus chicagoensis*, the Chicago chanterelle, which was characterized by Patrick Leacock, president of the Illinois Mycological Association

chio (G): Snow / white as snow. *Tyromyces chiorneus* (kye OR knee us) is the 'white cheese polypore', and is indeed as white as snow.

chlor (G): Green, greenish-yellow. *Russula chloroides* (klor OID ees) has a light greenish-blue band where the gills meet the stem (http://www.first-nature.com/fungi/russula-chloroides.php). Additionally, some members of the genus *Chlorophyllum* have green gills, at least at maturity.

chondr (G): Cartilage. For example, species with the epithet *chondroderma* (con dro DERM muh) have a cartilage-like skin (surface). Related is the taxonomic class of sharks, Chondrichthyes (cartilaginous fish).

chro / chroa / chroma (G): Color. *Chroogomphus* (kroo GOM phus) literally means 'colored spikes', but their common name is 'pine-spikes'. *Pholiota polychroa* (poly CROW uh) is an interesting multi-colored mushroom. Further, it is also used in *Russula heliochroma* (heal ee oh CHRO muh), which has the color and beauty of the sun.

chrysos (G): Gold / golden colored. *Clathrus chrysomycelius* (CRY so my SEAL ee us) has golden colored rhizomorphs (root-like aggregates of hyphae) at the base. 'Chrys' is a common word element of mushroom species. Can you think of any that use it?

chytra (G): A cooking pot or flower pot. Chytridiomycota (chy trid ee oh my COE tuh) is a taxonomic phylum of saprobic fungi whose zoospores develop in flower-pot shaped zoosporangia. 'Chytridiomycosis' is a deadly infection in amphibians caused by the chytrid *Batrachochytrium*.

cibar (L): Suitable for food. *Canathrellus cibarius* (sih BAR ee us) is a commonly sought after edible, and it can sometimes be found in your local grocery store.

cicada (L): A tree cricket, cicadas. *Ophiocordyceps cicadicola* (sih cade ih COLE uh) grows on cicadas.

cimic (L): A bug. Epithets *cimicarius* and *cimiciodorum* both indicate that the mushroom smells like bugs. The epithet *cimicifugatum*, used in a number of plant species, indicates that the plant odor repels bugs.

cincinnatus (CIN cin NAT us): Named after Cincinnati, Ohio, where species with this epithet were originally collected. *Laetiporus cincinnatus* is the common white pored chicken of the woods. It is found on Oak trees in the eastern United States.

cinctus (L): Banded. Used in *Russula cincta* (SINC tah) and *Russula citrinocincta*, which has yellow bands. 'Cincta' and related forms appears around the kingdom in several mushroom names.

cinereus (L): Ash-colored. *Cantharellus cinereus* (sci NARE ee us) is called the 'ash colored chanterelle' or 'ashen chanterelle'.

cinnabar (G): A red pigment. *Phallus cinnabarinus* (SIN nah bar RHE nus) literally means 'the reddish penis mushroom'. Several other mushrooms have this species name, can you think of one?

cinnamome / cinnamomeus (L) Cinnamon colored. *Coltricia cinnamomea* (sin nuh moe ME ah) is brown to reddish-brown, or cinnamon-colored.

circinan (L): To make round. The cap shape of *Cudonia circinans* (sir CIN ans) is generally rounded (circular).

cirrus (L): A curl. The base of *Collybia cirrhata* (cir RHA tuh) is twisted. The epithet is also used in *Clavaria* and *Hypoxylon.* In related terms, 'cirrus clouds' are curled.

citrinus (L): Lemon-colored. The outside of *Scleroderma citrinum* (sih TRIN num) is yellowish in color.

clad (G): Branched. *Marasmius cladophyllus* (clad OPH fill us) has anastomose (interlinked/branched) gills. A clade, a term often used in mycology, is a group of related species.

clarus (L): Bright, clear. The cap of *Russula claroflava* (klar oh FLAV uh) is bright yellow. A related word is 'clarity', which means 'something made clear'.

clast (G): Broken. *Coprinopsis clastophylla* develops broken gills. 'Clastic' rocks are comprised of broken pieces of other rocks.

clathr (L): Latticed, lattice work. *Hericium clathroides* (clath ROY dees) literally means 'having a likeness to lattice work'. You may also be familiar with the 'latticed stinkhorn', *Clathrus ruber.*

claud (L): Lame. *Claudopus* (CLOW doe pus) literally means 'lame foot' because their stems are somewhere between short and nearly absent.

Claussen, Peter (1877 – 1959): Peter was a German mycologist, a director of the Botanic Gardens in Marburg, and a long-time president of the German Botanical Society. In 1923, the genus *Claussenomyces* (clau sen oh MY cees) was named after him.

clava (L): A club. *Ophiocordyceps clavulata* (clav you LAT ah) is club shaped. The whole name literally means 'a snake-shaped club that is shaped like a little club'. Many species with a club-shaped stem have the epithet *clavipes.*

clitos (G): Slope. *Russula clitocyboides* (cly tos ih BOY dees) and the entire genus of *Clitocybe* have a somewhat sloped-form.

clype (L): A shield. The specific epithets *clypeatus* and *clypeata* (clip ee AH tuh) are used in nearly 100 species. A related word is 'clypeate', which means 'shield-shaped'.

coalitus (L): United, to join. The bases of *Hydnellum coalitum* are fused together. A related word includes 'coalition'.

coccin (L): Scarlet, red. *Cordyceps coccinea* (COE sin EE ah), to name one example, is red.

coccolobae (G): From *Coccoloba*, a Genus of flowering plants, which means lobed-seeds. It is used in a newly described chanterelle from Florida that Mary Smiley discovered and helped characterize, *Cantharellus coccolobae* (koe koe LOE bay). It grows in association with *Coccolob*a species, including *C. uvifera* and *diversifolia*, and I have heard that it is edible.

coccora (E): Thought to be short for 'cocoon', according to Arora, 2013. *Amanita calyptroderma* is locally referred to in California and other parts of the west coast of the United States as just 'coccora', a presumed reference to the cocoon-like universal veil.

cochlear (L): A snail's shell. *Ganoderma cochlear* (COE klee er) is shaped like a snail's shell. A related word is 'cochlea', the spiral-shaped inner ear.

coco (L): Coconut. *Lactarius cocosiolens* smells like coconuts, while *Botryosphaeria cocogena* is a parasite on coconut leaves. The epithet *cocoicola* indicates growth on coconut trees, but I do wonder what soft drink the authors of those species were drinking.

coelestin (L): Heavenly, represented by a sky-blue color. *Entoloma coelestinum* is sky-blue. On the other hand, *Inocybe coelestium* is hallucinogenic and was named after the gods of the heavens.

coffea (L): Coffee, the color of coffee, the color of the coffee plant's flower. *Ganoderma coffeatum* (COFF ey AT um) (currently *Humphreya coffeata*) is coffee-colored. What are the odds someone was drinking coffee when they were describing this one?

cognat (L): Related to. Species with epithets *cognata* (cog NAT uh) and *cognatus* were generally similar to other species. A 'cognate' word is one that is derived from the same root as another.

Cogniaux, Alfred (1841 - 1916) was a highly published and well-liked Belgian Botanist. Cogniaux was honored in *Cogniauxia*, the genus name of a cucumber plant, and then in *Neocogniauxi*a, an orchid. Kuntze wanted to recognize Cogniaux in a genus of fungi and didn't want to be redundant with the name of the cucumber or orchid, so he came up with *Biscogniauxia*, which means 'second Cogniaux'. The articles "*Biscogniauxia repanda*" (Omphalina vol 4, no 1, 2013 by Malloch) and "Who's in a name? *Biscogniauxia atropunctata*" (NJMA News, 46, 2, 2016, by Dawson) both give a much more extensive account of Alfred Cogniaux and *Biscogniauxia* than presented here, as well as pronunciations of the genus name. I crowd sourced the pronunciation with a group of Belgians, and support the pronunciation given by Malloch and Dawson for the genus name (beece cone YO zee uh). *Biscogniauxia atropunctata* and *B. mediterranea* are treated in Ascomycete Fungi of North America by Beug, Bessette, & Bessette (2014).

cohaer (L): Sticking together, coherent. *Marasmius cohaerens* (co HAIR ens) grows in a fused mass of fruitbodies.

Coker, William Chambers (1872 – 1953): Coker was an American mycologist and botanist. He authored the classics 'The Gasteromycetes of the Eastern United States and Canada' in 1928 (there is a copy on Amazon for just $4.35 right now!) and 'The Club and Coral Mushrooms (Clavarias) of the United States and Canada' in 1923 (there is a 1974 reprint for $5.19). He also co-authored books on plants and trees of the Carolinas. There are over 30 species named after him, including *Amanita cokeri* (CO ker eye), *Rhizopogon cokeri*, and *Psilocybe cokeriana.*

cola (L): An inhabitant. *Agaricus sylvicola* (sil VIC oh luh) grows in the woods, and *Boletus pinicola* grows among the pines. There are hundreds of other 'icola' species out there.

collinit (L): Smudged, covered in slime. *Cortinarius* and *Suillus collinitus* (col LIN ih tus), as well as other species with the epithet, are varying degrees of slimy.

collis (L): A hill: 'collinus' means hill-loving, and 'colliculus' means small hill. Common epithets include *collinus* (col LIE nus), *collina*, and *colliculosa* to name a few. For example, *Marasmius collinus* was named for its occurrence on hillsides. In related terminology, the superior and inferior colliculi are structures in the brain that look like little bumps (hills) protruding off the midbrain. Also, 'Collatina' (= Collina) is the Roman goddess of the hills.

collyb (G): A small coin. The caps of members of the genus *Collybia* (col LIB ee uh) can look like small coins from above.

color (L): Color. The epithet of *Phallus multicolor* (MUL tee color) literally means 'many colors'.

coltrice (Italian): Bed, couch. *Coltricia* (cole TRIS ee uh) is sometimes called 'Fairy's stool', which certainly sounds better than 'Fairy's couch'.

colum (L): A colander. The rays of *Myriostoma* (=*Lycoperdon*) *coliforme* (col ih FOR may) look like an upside down colander.

columba (L): Dove or pigeon, the genus name of the dove. *Russula columbicolor* (coe LUM bih color) is colored like a dove. In related terminology, the columbine flower is also named after a dove. Apparently, the five upside-down clustered flowers looks like five doves (try to imagine that one next time you see the flower). Also, 'Columba' is the 'Dove constellation'. However, R. *columbiana* was named after its discovery in Columbia.

colus (L): A distaff, a wool-wrapped spindle. *Cortinarius colus* has a fibrous stem, and *Pseudocolus* (originally just *Colus*), shares a likeness to a distaff. Ironically, *Pseudocolus fusiformis* literally means 'spindle-shaped false-spindle'.

com / compact (L): Together, joined. Russula section Compactae (come PAC tay) represents species that have firm and compact fruitbodies. Related words include 'communal' and 'compound'.

comatus (L): Hairy. *Craterellus luteocomus* (lue tee oh COME us) can have yellow (luteo) fibers (hair) on the cap. You may be more familiar with another mushroom that comes from the same word, *Coprinus comatus* (the shaggy mane, or lawyer's wig).

comeden (L): To destroy, consume. *Vuilleminia comedens* (COM eh dens) destroys bark.

communis (L): Growing together, common. *Schizophyllum commune* (com YOU ney) grows in clusters and has a very wide distribution. Related words include 'community' and 'commune'.

complicat (L): From *com*, meaning 'together', and *plicare*, meaning 'to fold'. Together, it means folded back over itself. The margin of *Stereum complicatum* is folded back on itself, which is complicated indeed. A related word is 'plicate', which means 'folded', and *Parasola plicatilis* is the pleated (folded) parasol.

compres (L): Squeezed together, straight. There are hundreds of species with the epithets like *compressa*, *compressum*, and *compressus*. Species like *Helvella compressa* have a straight stem. Other species are either compressed or straight (or maybe both).

compt (L): Band. Species with epithets *compta* and *comptus* have a band.

comptul (L): Luxuriously decked. *Agaricus comptulus* (comp TUL us) has a silky luster and *Psilocybe compta* is ornamented with shining spots.

con (L): The same. The 146 species with the epithet *concolor* have a uniform color. A related word is 'conspecific', a member of the same species.

concha (L/G): Shell. *Concha* (CON cuh), *conchata*, and *conchatum* are epithets that have been used in about 100 species and generally refer to the shell-like shape of a fruitbody, like *Poronidulus conchifer* (formerly *Trametes conchifer*). Conch piercings go through the 'concha', the shell-shaped part of the pinna.

concolor (L): Of the same color. *Russula concolor* (CON color) has a uniform color.

concrescens (L): Coalescing. The Bases of *Hydnellum concrescens* (con CRESS ens) are fused together, or conjoined.

condens (L): Pressed together, condensed. Epithets like *condensata* are used in nearly 100 species and usually indicates a clustering of fruiting bodies.

confertus (L): Crowded. *Lachnella, Nectria, Psathyrella*, and *Pezizella conferta* (con FER ta) have dense gills or a generally dense growth pattern.

confluere (L): To come together. *Gymnopus* and *Albatrellus confluens* (con FLU ens) grow in clusters with several fruitbodies that join together at the base. In related terminology, a 'confluence' is a joining of two rivers (like the confluence of the Mississippi and Missouri river).

confrag (L): Rough. Used in the epithets *confragosa* (con frag OH suh) and *confragosus. Daedaleopsis* and *Tubaria confragosa* are two of the more common applications of the word.

conic (G): Cone shaped. *Hygrocybe conicus* (CONE ih cus) is conical.

conis / konis (G): Dust. The spore print from *Pholiota conissans* (CON is sans) is abundant and appears like dust. Also, the 'konicellular layer' of the lateral geniculate nucleus of the thalamus in the brain appears like dust when it is stained.

contort (L): Twist together. The stem of *Clavaria contorta* is twisted.

convexus (L): Arched, convex. *Pulvinula convexella* (con vex EL luh) is a convex-shaped ascomycete.

convolvere (L): Rolled together, coiled pattern. *Gyromitra* and *Peziza convoluta* (con voe LOO tuh) have a convoluted growth pattern. In related words, a convoluted explanation is complex, twisted, and often confusing.

copiae (L): Plenty. *Craterellus cornucopioides* (COR nu cope ee OYE dees) literally means 'a little cup/crater that looks like a horn of plenty', and it is often found in great bounty. This mushroom is commonly called 'the black trumpet'.

copr (G): Dung. *Coprinus* (coe PRY nuss) literally means 'belonging to dung'. Species using this root are found on dung, compost, and in other Nitrogen-rich areas. They include the genera *Coprinus, Coprinellus*, and *Coprinopsis*, and the best known species utilizing this word is *Coprinus comatus*.

cor (L): The heart. *Coprinopsis cordispora* (cor dih SPOR uh) has heart-shaped spores. However, the epithet *cordobensis* means 'from Cordoba' (Spain), which itself was named after a person, not a heart.

corall (G): Coral. The epithet of *Hericium coralloides* (cor ull OY dees) literally means 'having a likeness to coral', and is indeed one of the 'coral mushrooms'.

cordyl (G): A cudgel, bump, or swelling, essentially, club-shaped. Members of the genus *Cordyceps* (COR dih ceps) are club shaped.

corem (G): Broom. The genus *Xylocoremium* (xy lo cor EM ee um) essentially means 'a broom on wood'. 'Coremoid' can be an adjective meaning 'broom-like'.

corium (L): Leather. *Hygrocybe russocoriacea* (rus so cor ee AH see uh) smells like Russian leather.

cornu (L): Horn. *Craterellus cornucopioides* (COR nu cope ee OYE dees) literally means 'a little cup/crater that looks like a horn of plenty', and it is often found in great bounty. Related terminology includes *cornu ammonis*, or 'ram's horn'.

coronadensis (COR ohn ah DEN sis): Named after the Coronado National Forest. *Phellinus coronadensis* was first described from the Coronado National Forest in southern Arizona, growing in association with white pine.

coronilla (L): Crown, wreath, garland. *Stropharia coronilla* is called the 'Garland Stropharia' in the States and the 'Garland Roundhead' in the UK, due to the garland-like appearance of the ring on the stalk.

corrugis (L): Having folds or wrinkles. *Hymenochaete corrugata* (kor roo GAH ta) is wrinkly. Related terminology includes corrugated cardboard.

cortic (L): Bark. *Gymnopilus corticophilus* (cor tih CAH phil us) grows on bark. In related terminology, your cortex is the bark of your brain, the outer layer of your brain tissue.

cortina (L): A curtain. Members of the genus *Cortinarius* (cor tin AIR ee us) usually have a prominent cortina (Google a picture of them if you are not familiar with them).

corusc (L): Flashing, shining. *Cortinarius coruscans* is shiny when dry. 'Coruscant' is an adjective meaning 'glittering' or 'sparkling'.

corvus (L): Raven. *Mycena* and *Tephrocybe* (*Collybia*) *coracina* (cor uh SEE nuh) are the color of ravens.

coryn (G): Club, club-shaped bud. The genus *Ascocoryne* is an ascomycete that looks like a club-shaped bud.

cossus (L): A larvae, specifically the goat-moth. *Hygrophorus cossus* was named after the odor of the goat-moth. The odor of the mushroom is variably described in books as 'fragrant', 'odd and non-mushroomy', and 'like a caterpillar'. I haven't had the pleasure of smelling this one, but I look forward to it. I wouldn't mind finding a goat-moth either.

costat (L): Ribbed. *Clitocybe costata* (cos TAH tuh) is called the 'Ribbed Funnel'. The epithet is also used with *Morchella* and around 100 other species. Our rib bones are called 'costae'.

cothurn (G): A calf-high boot worn by ancient Greeks and Romans. *Amanita cothurnata* (co thur NAH ta) is the 'Booted Amanita', and it is also used in *Coprinopsis cothurnata* and *Suillus cothurnatus.*

coton (G): The olive. *Cortinarius cotoneus* is the color of a wild olive, as are other species with the epithets *cotonea* and *cotoneus.*

cotyl (G): Cup-shaped. Members of the genus *Cotylidia* (cot ee LID ee uh) are named after their flute-shaped fruitbodies. A cotyledon, the leaf of a seed, is cup-shaped at first, and plants with two leaves are called dicotyledons (dicots).

cranion (G): Skull. *Calvatia craniiformis* (CRAY knee ih FOR mis) literally means 'the bald puffball shaped like a skull'.

crassus (L): Thick. *Gymnopilus crassipes* (CRASS ih peas) has a thick stem. There are examples in *Agaricus*, *Morchella*, *Boletus*, and many more. *Gymnopilus crassitunicatus* has a thick tunicate.

crater (G): A cup or the mouth of a volcano. *Craterellus* (cray ter EL us) species are generally vase-shaped, and the genus name literally means 'little cups'.

creber (L): Thick, crowded, frequent. *Ganoderma crebrostriatum* (CREE bro stri AY tum) has thick or crowded grooves.

creme (L): Cream. *Russula cremeoavellanea* (crem EH oh av ell an EE ah) is called the 'Cream Russula', a good thing the common name is not based on the second part of the epithet, because that would be confusing.

crepid (G): A shoe, more specifically a half-boot, a slipper worn by men. Used in the genus *Crepidotus* (crep ih DOE tus), but the half-boot is not the first thing I think of when I see a *Crepidotus.*

creta (L): Chalk. *Russula cretata* (creh TAT us) is chalk-colored.

cretac (L): Chalky. *Leucocoprinus cretaceus* (creh TAY cee us) is a chalky color. The 'cretaceous period' is characterized by extensive beds of chalk.

crinis (L): Hair. *Ophiocordyceps crinalis* (cry NAL iss) looks like it is covered with fine hairs.

crisp (L): Curl, wrinkled, wavy. The interlaced arms of *Clathrus crispus* (KRIS pus) are wrinkled, while the lobes of the fruitbody of *Sparassis crispa* are wavy, not because they are crisp or brittle. You may be familiar with the adjective 'crispate', which means 'having a wavy or curly edge'.

crista (L): Crest. *Calvatia sporocristata* (SPOR oh cris TAH tah) is a newly described species from Costa Rica that has spores with crests on them.

croceus (G): Saffron-colored. *Inonotus endocrocinus* (en doe chro SIGH nuss) has a dark pileus and the context is a rich saffron. The immature specimen has a reddish-brown and yellow surface.

crocodil (L): After the crocodile. *Agaricus* and *Cercopemyces crocodilinus* (croc oh dil EYE nus) each have a scaly pileus.

crucibul (L): An earthen pot or vessel. The genus *Crucibulum* (crew SIB you lum) that represents the group of bird's nest fungi shares a likeness to a crucible. Arthur Miller's play 'The Crucible' is derived from the other meaning of the word: a test or trial.

cruent / cruenta (L): To make bloody. *Hymenochaete cruenta* (crew EN tuh) is bright to blood red. *Hydnellum crentum* has blood-red droplets like the more commonly known *Hydnellum peckii.*

crust (L): Hard, typically a covering. *Hymenochaete crustacea* (crus TAY see uh) has a hard shell-like surface. In related words, 'crustaceans' are a group of arthropods with hard shells, and the 'crust' of the earth is defined as the outermost hard shell.

crustulum (L): A small pie. The pileus of *Hebeloma crustuliniforme* (cru stu lin ih FOR may) is shaped like a pie. Indeed, the common name of this mushroom is 'Poison pie'.

crypt (G): Vault, hidden. *Cryptoporus* (crip toe POR us) has hidden pores, and *Clitocybe cryptarum* can grow in cellars. Related words include 'crypt', 'cryptic', 'cryptogram', 'decrypted', and 'encryption'.

cubensis (L): From Cuba. *Psilocybe cubensis* (cue BEN sis), perhaps the most commonly consumed hallucinogenic mushroom in the world, was originally described from Cuba.

cucum (L): Cucumber. *Ophiocordyceps cucumispora* (cue cum ih SPOR ah) has cucumber-shaped spores.

cula / culus / culum (L): A suffix added to words to indicate a diminutive form of the word. It is used is many species like *Auricularia* (little ear), *Craterellus calyculus* (little cup), *Ganoderma tuberculosum* (little bump), *Collybia acicularis* (little point), and others.

culm (L): Stem or straw. A total of 68 species have had the epithet *culmicola*, indicating growth on straw or a stalk in general.

cumul (L): A heap or mass. *Hydnellum cumulatum* (cue mu LAH tum) forms a fused mass (somewhat like *H. concrescens* and *H. coalitum*). In related terminology, you have probably heard of 'cumulonimbus' or 'altocumulus' clouds.

cuneus (L): A wedge. *Tricholoma cuneiforme* (cue nay ih FOR meh) has a wedge-shaped fruitbody, *Dermoloma cuneifolium* has wedge-shaped gills, and *Lepiota cuneatospora* has wedge-shaped spores. In the brain, the 'cuneus' is a wedge-shaped portion of the occipital lobe.

cuniculus (L): Underground passage, tunnel. *Agaricus cuniculicola* (cue nic you LIC oh luh) grows along the tunnels of underground mammals.

cuph / cyph (G): Curved. *Cuphophyllus* was named after its curved gills.

cupressus (L): Cypress. *Fomitiporia cupressicola* (cue press ih COLE uh) grows on cypress trees.

cuprum (L): Copper. *Ganoderma cupreolaccatum* (coo prey oh lah CAH tum) is copper colored.

Curtis, Moses Ashley (1808-1872): Curtis devoted much of his later life to mycology. *Lycoperdon curtisii* was named after him.

curtus (L): Short. *Russula curtipes* (CUR tih pees) has a short stem. It is also used in *R. curtispora*, which has short spores. In related terminology, someone who is curt is rudely short in their dialogue.

cusp / cuspis (L): Pointed. *Phellinus bicuspidatus* (bi cuss pid AT us) is named because the setae have two points. In related terminology, you have cuspid (having one point) and bicuspid (having two points) teeth.

cutis (L): Skin. *Leucopaxillus cutefractus* (cue teh FRAC tus) might be cute, but the epithet indicates that the skin of the pileus cracks. Related words include 'cutaneous' and 'cuticle'.

cyan (G): Dark blue. *Elaphomyces cyanosporus* (cye an oh SPOR us) has dark blue spores.

cyath (G): A cup. *Cyathus* (sigh ATH us) *striatus* is shaped like a cup, and is commonly called 'the bird's nest fungus'.

cyb (G): Head. *Dermocybe* (der MAH suh bee) literally means 'skin head', a reference to the dry cap.

cycl (G): Circle. *Russula cyclosperma* (cy clo SPERM uh) has nice round spores.

cylichn (G): A small cup or little box. *Ascocoryne cylichnium* (cy LICH nee um) is shaped like a little cup.

cyphell (G): The hollow of the ears. *Hohenbuehelia cyphelliformis* (cy phel lih FOR mis) and *Lachnellula cyphelloides* resemble the inside of the ear.

cyttar (G): A partition, generally, a honeycomb, because it has so many partitions. Used in the genus *Cyttaria* (cih TAR ee uh) and specific epithets like *cyttarioides*.

D

dacry (G): Tears. *Dacrymyces* (dah crah MY sees) and *Dacryopinax* are both reminiscent of tears.

daedalus (G): In Greek mythology, Daedalus was a craftsman who made the labyrinth that would hold the minotaur. It is used in the genera *Daedalea* and *Daedaleopsis*. *Daedalea quercina* has an intricate maze-like pore surface, and is named to pay homage to Daedalus. Pronunciations are all over the map, but *Daedalus* should be pronounced 'DEAD ul us' or 'DEED ul us', But I've heard it pronounced 'dead AL ee us'.

dal (G): Firebrand (charred wood). *Daldinia* (dal DIN ee uh) looks like charred balls. The British common name of *Daldinia*, 'King Alfred Cakes', is also descriptive. The king left his bread (cakes) in the oven too long and they came out looking like charred remnants, similar to *Daldinia concentrica*.

dama (L): Fallow-Deer. *Xylaria* and *Podostroma cornu-damae* (cor nu – DAH may) are shaped like the antlers of the fallow-deer, *Dama dama*.

Darwin, Charles (1809 – 1882): Charles Darwin made a collection of a fungus in 1832 on his famous trip on the HMS Beagle in Tierra del Fuego. The fungus was locally known to be edible, and Darwin tried it: "It has a mucilaginous, slightly sweet taste, with a faint smell like that of a mushroom". He brought the collection back to England and a decade later, Berkeley named it *Cyttaria darwinii* (dar WIN ee eye) after him.

dasy (G): Hairy, shaggy, dense. *Pholiotina dasypus* has a hairy stem. There are also a number of fungal species named after their association with the plant *Dasylirion*, which itself means 'hairy-lily'.

datronia (completely made up): The genus name *Datronia* is merely an anagram of *Antrodia:* courtesy of Donk.

dauc (G): A carrot, genus *Daucus.* The base of the stem of *Amanita daucipes* (DAU cih peas) is carrot-shaped.

dealbat (L): White, made white, white-washed. *Clitocybe dealbata* (day al BAH tah) is a white mushroom.

debil (L): Weak, crippled. *Mycena debilis* (de BIL is) and many others with this epithet are weak and fragile. A related word is 'debilitate'.

deca (G): Ten. *Lyophyllum decastes* (day CAS tees) grows in large clusters of around 10 fruitbodies.

deceptiv (L): Deceptive. Forty-six species have had the epithets *deceptiva* or *deceptivus*, typically due to subtle differences between them and a similar species.

deciduous (dee SIDGE you us) **warts**: Warts on the exoperidium of many puffball species can be scraped off (or otherwise fall off with age), and these are called deciduous warts. They are common in *Lycoperdon perlatum*, which also has textbook pyramidal warts.

decolor (L): Faded. Decolorantes (de color AHN tays), one of the seven sections of the Russulas, has specimens whose context grays or blackens. Several specific epithets utilize *decolorans.*

decorus (L): Elegant, decorative, decorated. The cap of *Pholiota decorata* (de cor AH ta) (formerly *Gymnopilus decoratus*) appears to be decorated - it has a streaky-fibrillose to scaly cap.

delect (L): Delightful. Species like *Inocybe delecta* and *Spongipellis delectans* (de LEC tans) are pleasing. A related word is 'delectable'.

delicatus (L): Delicate, pleasuring. *Mycena* and *Russula delica* (DELL ih cuh) are delicate.

delph (G): Brother. The genus *Clavariadelphus* (cluh var ee uh DEL fus) is so similar to the genus *Clavaria* that they could be brothers.

demiss (L): Fallen, drooping. *Amanita* and *Omphalina demissa* (day MISS uh) appear to be droopy.

densus (L): Dense. *Ganoderma densizonatum* (den si zoe NAY tum) has dense zones.

depend (L): Hanging down. The fruitbody of *Coltriciella dependens* (dee PEN dens) often hangs down.

derm (G): Skin. The 'derm' in *Dermocybe* (der MAH suh bee) refers to the skin of the mushroom. *Dermocybe* specimens are more brightly pigmented than your average *Cortinarius*, and are used to create dyes. Related words in our language include 'dermis' and 'dermatologist'.

des (L): Not. *Armillaria* means 'ring' or 'band', and most examples in the genus have a pronounced annulus. However, we have *Armillaria tabescens*, the so-called 'ringless honey mushroom', which never made much sense given the meaning of *Armillaria*. Problem solved! *Desarmillaria*, elevated from subgenus to genus in 2017, literally means 'not an Armillaria', 'not ring', or 'ringless', and is used with the ring-less honey mushrooms. *Armillaria tabescens* is now formally *Desarmillaria tabescens*. Before you join the chorus of grumblings over the name change, remember that we now have a descriptive name (*Desarmillarira*) to use over the contradiction of using '*Armillaria*' and 'Ringless' in the same sentence.

destruo (L): To destroy. *Phaeoramularia destruens* (des TRUE ens) kills its host.

detrit (L): Waste, debris, to wear away. Used in many species including *Galerina detriticola* (deh tri tih COLE uh), which grows on root debris. A related word includes 'detriment', a state of being harmed, damaged, or at a loss. Also, in biology, 'detritus' is dead organic material like leaf litter.

deust (L): Burned up. *Kretzschmaria deusta* (day OOHST uh) (formerly *Ustulina deusta*) looks like a burned up mass.

diabolic (G): Devilish, fiendish. The original description and placement of *Cortinarius diabolicus* (di uh BOL ick us) was questionable. Unsurprisingly, most species with this epithet predated the modern genetic revolution and have been renamed. However, it should be noted that some species with this epithet are just reddish.

diaphan (G): Transparent, diaphanous. *Cotylidia diaphana* (di af AN uh) can be transparent along the edges.

dicty (G): A net. *Scleroderma dictyosporum* (DIC tee oh SPORE um) refers to its net-like, or reticulate spores. You may also be familiar with the genus *Dictyophora*, the latticed stinkhorn, which have an elaborate net.

didym (G): Two-fold, doubled. *Agaricus didymus* (did EE mus) was named because it is very similar (is a twin with) to another species, *A. gemellatus*. Related, the 'epididymis' is a duct that goes into each testicle.

dilat (L): Dilated, spread out. *Clitocybe dilatata* (die luh TAH tah) has an enlarged stipe at the base, and is often fused with others to form a large clump.

dilut (L): Watery, diluted. *Cortinarius dilutus* (die LOO tus) has muted colors.

diospyros (G): Fire of Zeus, the genus name of the persimmon tree. *Agaricus diospyros* stains a persimmon color when rubbed. Fungi with epithet *diospyricola* grow on *Diospyros* trees.

dipl (G): Double. The epithet *diplomorpha* (dip loh MOR phuh) suggests two different forms, and *diplospora* refers to two spores. In related terms, 'diplopia' is double vision.

disckos (G): A disc. *Russula nigrodisca* (nigh grow DISC uh) looks like it has a black disc on the cap.

disjunct (L): Not joined, or separated. Examined specimens of *Gymnopus disjunctus* were from geographically distinct areas.

dispers (L): Scattered, dispersed. A whopping 443 species names utilize some form of dispersed in their names. In *Hypholoma dispersum* (dis PER sum), specimens are physically dispersed and generally grow solitary. The first part of the word here, 'dis', specifically means 'widely', and this is utilized in words like 'distance', 'distend', and 'discrepancy'.

dissem (L): Scattered. *Coprinellus disseminatus* (dis sem in AT us) grows in large troops that are scattered about. Related to 'disseminate', to spread or distribute information.

domus (L): House. *Peziza domiciliana* (dom ih cil ee AN uh) is frequently found growing in basements. Related words include 'domestic' and 'domicile'.

dry (G): Oak. *Gymnopus dryophilus* (dry AH fill us) and *Strobilomyces dryophilus* grow under oaks (and the other half of the names, *philus*, means that they love the oak). The oak tree has a mycorrhizal relationship with many mushrooms, so it is no surprise that numerous mushrooms are named after the tree.

dryad (G): In Greek mythology, Dryad (DRY ad) was a tree nymph. Specifically, 'dry' is Greek for 'oak tree' (think *Gymnopus dryophilus*, which is found under oak trees), so Dryad was a nymph of oak trees. Dryad's saddle, the common name of *Cerioporus squamosus* (formerly *Polyporus squamosus*), grows on boxelder, silver maple, elm, aspen, chestnut, ash, beech, and yes, sometimes even oak trees.

Dudley, William Russel (1849 - 1911) was a Botanist who went to college at Cornell University but ended up at Stanford University where his collections became the Dudley Herbarium. In 1888, he found an undescribed red cup fungus in Ithaca, New York, and sent it to his friend Charles Peck. Six years later, that specimen would be referred to as *Peziza dudleyi*, and later renamed as *Sarcoscypha dudleyi.*

dulcis (L): Sweet. *Russula dulcis* (DUL cis) must be a sweet-tasting mushroom, although I have not tasted it.

duplic (L): Double, duplicate. *Phallus duplicatus* (dupe lih CAH tus) literally means 'the double penis'. I would say that this name takes the cake for the weirdest mushroom name. This fungus has a skirt that is likely to adhere to the stem. I cannot find a definitive answer to what part of this mushroom is doubled, but I imagine it somehow refers to the skirt.

dur (L): Hard, strong, durable. *Russula dura* (DUR uh) must be more durable than most other Russulas. In related terminology, Duracell batteries are named because they are durable, and the *dura mater* (literally 'tough mother) is a very tough layer that surrounds and protects the brain and spinal cord.

E

earl (G): Referring to the springtime. *Gymnopus earleae* (ER lee ay) fruits in the spring, which is early for a mushroom.

eburnea / eburneus (L): ivory (color). *Arrhenia eburnea* (ee BURN ee uh) and *Hygrophorus eburneus* are both ivory colored. This root word is used in over 100 specific epithets.

eccent (L): Off-centered. The stem of *Clitocybe eccentrica* (ee CEN tric uh) is off-centered. This was noted as an eccentric (strange) feature of the mushroom.

eccilia (G): Hollowed out. The stem *Eccilia* species are hollow.

echin (G): A hedgehog, and members of the phylum Echinoderm, including sea urchins, sea cucumbers, sea stars, and sand dollars. In the mushroom world, members of the genus *Echinoderma* (ee kine oh DER muh) are Lepiota-like mushrooms that have scaly caps.

edodes (L): Edible. *Lentinula edodes* (ee DOE dees) is edible, and actually accounts for 10% of the world's cultivated mushrooms.

edul (L): Edible. The name *Boletus edulis* (ED jul is) literally means 'a superior kind of edible mushroom'.

effusus (L): Spread out. *Exidiopsis effusa* (ef FEW suh) looks like white peanut butter spread on a limb with a butter knife.

elach (G): Small. *Cortinarius elachus* (ee LACH us) and several other species with this epithet are small.

elaph (G): Deer. *Elaphomyces* (ee LAPH oh MY cees) *granulatus* is called the 'deer truffle'. *Elaphocordyceps* species are related to the *Cordyceps* genus.

elast (L): Stretchy, elastic. *Helvella elastica* (ee LAS tic uh) has a flexible stem.

elat (L): High, raised, lofty. *Hydnum elatum* (ee LAT um) is tall. A related word is 'elated', meaning 'ecstatically happy'.

embol (G): A wedge. *Galerina embolus* (EM bo lus) has wedge-shaped gills. 'Embolus' is a synonym of the Latin 'cuneus', and a related word is 'embolism'.

emetos (G): Vomiting. *Russula emetica* (ee MET ic uh) can cause gastrointestinal distress. *Russula emeticicolor* is the color of vomit: it is rosy red with violet tints.

enelensis (L): The Latinized form of 'NL', the provincial abbreviation of Newfoundland and Labrador. *Cantharellus enelensis* was originally described from this province, but its range is not restricted to it.

enoki (Japanese): Nettle tree, Chinese hackberry. Enoki (cultivated *Flammulina velutipes*) is named after the Chinese hackberry it can grow on.

ensis / ense (L): Belonging to. *Clathrus oahuensis* (oh ah hoo EN sis) is a species from Hawaii, while *Clathrus xiningensis* was originally described from China. Species with this suffix can be from a geographical region or a specific habitat. For example, *Agaricus arvensis* belongs to a field. Hundreds of species are named after their location or habitat – which one comes to your mind?

enteron (G): Intestine. *Auricularia mesenterica* (mes en TER ic uh) is a wood-ear jelly fungus that looks like intestines. *Tremella mesenterica*, commonly called *Witch's Butter*, also looks like a clump of intestines, but 'Witch's Butter' sounds more appealing than 'Witch's Intestines'. In related terminology, the 'enteric nervous system' is part of your peripheral nervous system that controls your gastrointestinal tract, and the 'mesentery' is a fold of tissue that connects to the intestine.

entomopathogenic fungi: A fungus that infects and frequently kills insects. The most well-known examples are *Cordyceps* and *Ophiocordyceps* species, and these infect insects like caterpillars, wasps, and ants. Another entomopathogenic fungus, *Pandora neoaphidis*, kills Peach Aphids, and is commercially used to control the pest.

entos (G): Inner. *Entoloma* (en tah LOW muh) literally means 'inner fringe', a reference to the inrolled margin of the pileus.

ephem (G): Brief, lasting a day. *Coprinellus ephemerus* (ee FEM er us) and 26 other species that use this epithet don't last very long after they fruit. 'Ephemeral' is also used frequently as an adjective to describe such a mushroom.

ephip (G): Literally 'upon a horse', generally a saddle. *Helvella ephippium* is shaped like a saddle. In *Rhizophydium ephippium*, the sporangium is saddle-shaped.

epi (G): Upon. *Mycina epichloe* grows on grass, and *M. epiphyllus* grows on leaves. An 'epiphyte' grows on other plants.

equestre (L): Horse, one who rides horses. Common names of *Tricholoma equestre* include 'man on horseback' and 'saddle-shaped Tricholoma'. Different sources stress the saddle-shape of the cap or the "distinguished appearance in the woods" (ME Hard, 1908) when discussing the etymology.

ereb (G): The place of nether darkness, and as far as mycology is concerned, just dark. *Hebeloma*, *Cyclocybe*, and *Thelephora erebia* (er EB ee uh) are all dark. In Greek mythology, 'Erebus' is the personification of darkness, as well as the darkness on the way to Hades.

ergot (French): Spur. Used in ergot (air GOT), a common name for a group of fungi that includes the genus *Claviceps*. The sclerotia of these fungi look like a cock's spur. Some chemicals in these fungi are called ergot alkaloids, and can cause hallucinations, gangrene, or both.

erinaceus (L): A hedgehog. *Hericium erinaceus* (air in AY cee us) or (air in AY shus) is another common *Hericium* in North America. I have heard *erinaceus* pronounced the two ways above with equal frequency.

erio (G): Wool. *Lepiota* and *Amanita eriophora* look like they bear wool.

ermin (L): The stoat, a weasel-like animal. The epithets *erminea* (er MIN ee uh) and *ermineus* are used with a number of genera, including *Lactarius*,

Lepiota, and *Mycena*, and refers to the color of the stoat, which ranges from white to chestnut brown depending on the location and season.

erumpens (L): Breaking out. *Russula erumpens* (ee RUM pens) is called the 'erupting Russula'. The word 'erupt' is derived from this Latin word.

eryngii (L): After the host plant, *Eryngium*, which means 'sea holly'. *Pleurotus eryngii* (air IN gee eye) is parasitic and associated with the roots of the European *Eryngium* species. 'The King Oyster', as it is commonly called, is now a popular cultivated species.

erythros (G): Red. *Russula erythropus* (air RHEE thro puss) has a distinctly red stem. In related terminology, red blood cells are erythrocytes.

escens (L): Turning, changing, or becoming. *Amanita rubescens* (roo BES cens) turns reddish, *Psilocybe azurescens* stains blue, and *Agaricus urinascens* develops a urinous odor with age.

estre / estris / ester (L): Origin or place of growth. *Hericium alpestre* (al PESS tre) grows in the Alps.

eu (G): True, good, well. Used as a prefix in genus names, specific epithets, and in many other applications. It is used in 'eucalyptus' and 'eukaryotic', and thousands of other words.

evade (L): Escaping, evading. *Rhizopogon evadens* (ee VAY dens) typically has buried fruitbodies.

evern (G): Flourishing. *Cortinarius evernius* (ee VIR nee us) flourishes (but which mushroom doesn't under the right conditions?). Additionally, *Tremella everniae* is named after its host, *Evernia mesomorpha*, the 'ring lichen'.

evolvo (L): To unroll. *Cylindrobasidium evolvens* (ee VOLV ens) has new growth at its margins, as if the species is unrolling. Think 'evolve' and 'revolve', or 'volver' ('to return' in Spanish).

excavat (L): Hollowed. *Fomes excavatus* has a hollowed out hymenium. Excavate is a related word.

excels (L): Tall, elevated. *Craterellus excelsus* (ex CELS us) gets to be a whopping 6 inches high. I wish we had these in North America and that they were as common as black trumpets!

excoriat (L): To remove the skin, abrade. *Lepiota excoriata* (ex cor ee AH tuh) was named because of the way the scales appear to be moving inward from the margin of the cap. Related, 'excoriate' means 'to berate', literally 'to remove someone's skin'.

exidia (L): Staining, exuding. The genus *Exidia* (ex ID ee uh) is another genus within the Auriculariaceae, and some examples sometimes have an exudate.

exigu (L): Short, small. *Aleuria exigua* (ex IDGE you uh) is only 5 mm in diameter! A related word is 'exiguous', meaning 'very small'.

exil (L): Small, thin. *Inonotus exilisporus* (ex ill ih SPOR us) has slender spores. *Exilis* and *exile* are utilized in over 160 species names.

exim (L): Exceptional, uncommon. Used in *Auricularia, Peniophora, Lachnella,* and *Morchella eximia* (ex IM ee uh). Related, 'eximious' means distinguished.

expallens (L): Becoming pale. *Pseudoclitocybe expallens* (ex PAL lens) (formerly *Clitocybe expallens*) becomes pale with age.

exsud (L): To exude. The pores of members of the genus *Exsudoporus* exude yellow droplets when young.

F

fagus (L): Beech tree. *Gymnopilus fagicola* (fah gih COLE uh) grows on beech trees.

falcatus (L): Sickle-shaped, bent, curved, hooked. *Clavaria falcatispora* (fal cat ih SPOR uh) has sickle-shaped spores. In a related word, the falcon (genus *Falco*) got its name because of their hooked beaks.

fallax (L): Deceptive. Has anyone been deceived by *Craterellus fallax* (FAL ax). Anyone? That is what I though. Perhaps they were named because they are a little hard to find for the untrained eye? In France, these are called 'trumpets of the dead', because it is believed that they are little trumpets sticking out of the ground that get played at night from underground dead people (think about that the next time you are cooking some up!).

farina (L): Flour, meal. The stem of *Russula farinipes* (fair IN ih pees) is covered in a flour-like powder.

fascia (L): A band, in small bundles. *Hypholoma fasciculare* (fas sic you LAR ey) grows bundled in a small cluster. This root is used in several hundred species names. In related words, your brain has several 'fascicles' (bundles of nerve fibers), and 'myofascial release' utilizes the same word element too. Also, 'fascism' is a bundle or group of people with complete power.

fastig (L): Sloping up, an incline. Used in the now deprecated name *Inocybe fastigiata* (fas tidge ee AH tuh), where it refers to the shape of the cap, and in many other specific epithets too. This definition is confused with that of 'fascia' in at least one mycology dictionary.

fav (L): Honeycomb. The pore surface of *Favolus* and *Neofavolus* is shaped like a honeycomb.

fel (L): Bile. *Russula fellea* (FEL lee uh) and the widely known *Tylopilus felleus* are named after their taste – the gut wrenching taste of bile. The expression 'Bitter as Bile' ties this species to the Latin derivative.

fennic (L): Finland. *Ramaria fennica* (FEN nih cuh) was originally described from Finland, but it is also found in North America and Australia.

fer (L): To bear. It is also used as a suffix in many epithets, like *Ductifera* (duck TIF er uh) which has (bears) ducts, and in *Clitocybe morbifera*, which is toxic (bear's disease).

ferre (L): Iron, iron colored. The full name *Fuscoporia ferrea* (fare ee ah) means *dark brown-pored, iron-colored.*

festivus (L / Seinfeldism): Festive, gay, adorned with bright colors, related to a feast, and finally a secular holiday celebrated on December 23rd that denounces the commercialism of the season and gives people a chance to air their grievances. *Phaeocolllybia festiva* (FES tiv uh) is handsome. It is also used in *Clitocybe festiva*, *Austroboletus festivus*, and *Porphyrellus festivus.* I air my grievances at being unable to find why these three species are named 'festivus'.

fibul (L): Clasp, brooch, button. *Coriolus fibula* (FIB you luh) is button-shaped, and although *Rickenella fibula* is as cute as a button, the Simon & Schuster guide says it is named *fibula* because it is pin-shaped. The fibula, the lower leg bone, is named after its resemblance to a safety pin.

fidus (L): To split. Although deprecated, *Coprinus quadrifidus* (quad rih FID us) was named because of the tendency to split into four parts upon maturity. It is now called *Coprinopsis variegata*, but the old name is slow to die.

fil (L): Threadlike. *Ophiocordyceps filiformis* (fill ih FOR mis) is even more slender than the thin *O. gracilis.* Other related words include 'filament', 'filamentous', and 'filiform papillae' on the tongue.

filic (L): A fern. Several hundred fungi species allude to ferns. Many of the epithets denote growth on ferns, like *filicicola* (fil ic ih CO luh). Old taxonomic names for ferns include *Filices* and Filicophyta.

fim (L): Dung. *Paneolus* and *Cheilymenia fimicola* (fim IH cole uh) are just two examples that use this word element that indicates growth on dung.

fimbr (L): A fringe. *Geastrum fimbriatum* (FIM bree AY tum), the 'fringed earthstar', has fine fibers at the peristome (it has a fibrillose peristome).

fimet (L): A dungheap. About a hundred species that have the epithets *fimetaria* and *fimetarium* grow on a pile of dung. 'Fimetic' is an adjective to describe something that pertains to dung.

fiss (L): split. *Arrhenia fissa* (FISS uh), and nearly 200 other species, are split in some way. A related word is 'fissure', which can be a split in a rock or a deep groove in the brain (the central fissure separates the two hemispheres).

fistul (L): A tube, hollow, a pipe. This is used as an adjective to describe a hollow stipe. It is used in a number of species names, including *Clavaria*, *Ramaria*, and *Russula fistulosa* (fis tyou LOW suh). In related terminology, a bladder fistula is when there is an opening between the bladder and some other organ. A fistula may also be implanted to connect two organs.

flabellum (L): A small fan. *Ganoderma flabelliforme* (fluh bell ih FOR meh) is shaped like a small fan. Although this is a deprecated name, *flabell* gets hundreds of hits at Indexfungorum for both genera and specific epithets.

flaccid (L): Flabby, flaccid. The epithets *flaccida* (FLAH cid uh), *flaccescens* (ecoming flaccid), *flaccidum*, and *flaccidus* are used in over 150 mushroom species, but none from the Phallaceae.

flamm (L): Flame. *Flammulina* (flam myou LIE nuh) and *Flammula* are reminiscent of little flames. The epithets *flammeus* and *flammea* reference a flame-like color. Related words include 'flamboyant', 'flammable', 'flambeau', and even 'flamingo'.

flavus (L): Yellow. *Calvatia rubroflava* (RUBE rho FLA vah) is white, then yellow, then an orangish-red. The flesh also stains yellow.

Flett, J. B. studied, detailed, and photographed a mushroom in the 1930's that we now call *Albatrellus flettii* (FLET tee eye). His documentation can be seen in the 1941 publication in *Mycologia* by Elizabeth Morse (Vol 33, no. 5, pages 506-509), and should be an inspiration to us all. I can't find anything else about him, but his name lives on in 'Flett's Polypore' – or maybe it should be 'Flett's Albatrellus'.

flex (L): To bend. This obvious word element has been used in over 300 species names. *Cortinarius flexipes* (FLEX ih peas) has a flexible stem and *Peziza flexosa* is generally pliable.

flocc (L): Flock or tuft of wool. Members of the genus *Flocculaia* (floc cue LAR ee uh) have wooly remnants of the partial veil on the stipe.

fluere (L): To flow. The genus *Lactifluus* (lac tih FLU us) is characterized by flowing milk. 'Fluids', 'flue', and 'fluvial' are among some related terms.

foci / focus (L): Fireplace / hearth. *Coltricia focicola* (fahs ih COLE uh) grows in association with fire (charcoal, ashes, etc…). However, other *Coltricia* species grow in the same habitat, so this is not a unique identifying feature.

foen (L): Hay. *Aspergillus foeniculicola* (foy nic you LIC oh luh) or (foe nic you LIC oh luh) grows on grass. *Panaeolus foenisecii* is commonly called the 'lawn mower's mushroom', but is also called the 'hay mushroom'. I have heard more pronunciations of this species than any other species. Because of the *oe* diphthong, the beginning should pronounced (foy), but (pho) is probably more common, and 'proper' pronunciation is mostly driven by usage.

foetid / fetid (L): Foul-smelling. *Lycoperdon foetidum* (foe ET id um) is one stinky puffball. *Hymenogaster foetidus* is another stinky fungus.

folium (L): Leaf, gill. *Gymnopilus pulchrifolius* (pul chri FOLE ee us) has beautiful gills.

fomes/fomit (L): Tinder. *Fomes* was named after its historical use as a fire-starter. *Fomes fomentarius* is the 'Tinder Polypore'.

font (L): Fountain, a spring. *Galerina*, *Panaeolus*, and *Psathyrella fontinalis* all give reference to a fountain, and the brook trout is *Salvelinus fontinalis*. *Mycena fonticola* was first found near a spring. Related is the 'fontanel', the hollow portion of a baby's skull akin to a place where a spring might arise.

form (L): Shape. *Craterellus tubaeformis* (TOO bay FOR mis) has a stem that is shaped like a tube (it has a hollow stem). As far as common names, it is called the 'yellow-foot chanterelle' or 'winter chanterelle' and is a choice edible.

formosus (L): Finely formed or beautiful. *Cantharellus formosus* (for MOE sus) literally means 'the beautiful Chanterelle' and I can think of no better name for it. This species is the 'Pacific golden chanterelle'. There are 24 other *formosus* species on Indexfungorum.

fornix (L): Arch. *Geastrum fornicatum* (for nih CAY tum) is known as 'the arched earthstar'. It was originally described in *Fungus anthropomorphos* in 1671 (although some sources incorrectly state 1688). The word 'fornication' is also derived from 'fornix', but it has nothing to do with sexual positions. There was a famous archway (a fornix) in an ancient city where members of parliament found prostitutes. Thus, the word 'fornix' led to the word 'fornication'. The 'fornix' is also a band of arching white fibers in the brain.

frac (L): Break. *Elaphocordyceps fracta* (FRAC ta) easily breaks. In related words, you can think of a bone fracture, or even fracking, which is short for hydraulic fracturing.

fragaria (L): Strawberry. *Russula fragaricolor* (fruh GARE ih color) was named after the genus of the strawberry, *Fragaria*, itself named after *fragrum*, the Latin word for 'fragrant'. *Russula fragricolor* has a similar color as strawberries.

fragrare (L): Sweet smelling, fragrant. *Russula fragrans* (FRAY grans) is obviously fragrant, but I am including it due to prevalence: There are over 100 fungal species listed on Indexfungorum.com that use this word element. *Russula fragrans* and *R. fragrantissima* are two.

Franchet, Adrien René (1834 – 1900). Franchet was a French botanist and traveled far and wide on his expeditions. Many plant species are named after him, and in the fungal world, the European *Amanita franchetii* was named after him in 1889. *Amanita augusta* from the western United States used to be called *A. franchetii*, and this misapplied name is found in many older books and a few modern ones.

fraud (L): A cheat. *Hydnellum fraudulentum* (fraud you LEN tum) is a fraud! It masquerades and takes the identity of *H. caeruleum*, yet the only reward for its capture and proper identification is the realization of a job well done.

frax (L): The Ash tree, *Fraxinus*. There are over 450 hits on Indexfungorum for species named after the ash tree host, including *Perenniporia fraxinea* (frax in EE uh).

Fries, Elias Magnus (1794 – 1878): CG Lloyd devoted an entire issue of Mycological Notes (number 32) to Fries, including detailed family history and a synopsis of his primary contributions. Fries introduced many names of well-known mushrooms including *Schizophyllum commune*, *Marasmius oreades*, and Gasteromycetes just to name a few.

frio (L): Crumble, crumbly, friable. *Calvatia friabilis* (fry AB il iss) is friable. Try to use the word 'friable' in a sentence today.

frondis (L): A leaf. *Hydnellum frondosum* (fron DOE sum) looks like a bunch of leaves. Another species, *Grifola frondosa* (hen of the woods) also looks like a leaf cluster.

Frost, Charles Christopher (1805 - 1880). Frost described many new species of boletes, mostly in the northeastern United States. The frequently encountered *Boletus frostii* is named after him, as are around 40 others, which use the epithets *frostii* and *frostiana*. *Frostiella* was named after him as well.

fuc (G): The genus name of a seaweed. Nearly 20 species have epithets that indicate a seaweed-like fruiting body, like *Tremella* and *Xylaria fuciformis* (few cih FOR mis), none of which look like or were named by Karl Fuckel.

Fuckel, Karl (1821 - 1876) was a German botanist and mycologist who had an impressive list of publications. Fuckel (few kel) was a pharmacist at first, but he married into money, and after his wife died, he immediately retired and became a botanist, before finally turning to mycology. If anyone ever offers you a good *fuckeliana*, you should definitely take them up on it, as it is bound to be one for the memories, and one you'll probably want to document it with photographic evidence as well. There are numerous species with the epithets *fuckeliana*, *fuckelianum*, and *fuckelii* that all refer to him.

fuco (L): To paint or dye. *Tricholoma fucatum* (few CAT um) has a cap that appears dyed (at least according to WG Smith, British Basidiomycetes, 1908).

fulgens (L): Shining. *Caloscypha fulgens* (FUL gens) is brightly colored, as are other fungi with *fulgens* in their names.

fulig (L): Soot. The top of the pileus of *Lactarius fuliginosus* (fue lidge in OH sus) is soot colored (dingy ash or gray), and the slime mold *Fuligo septica* is yellow, but soot-colored when it dries.

fulmineus (L): Sparkling, brilliant. *Chroogomphus fulmineus* (ful min EE us) is a sparkling or brilliant mushroom.

fulv (L): Reddish-yellow, tawny, golden colored. Used in *Amanita fulva* (FUL vuh) and in other species to describe color.

fum (L): Smoke, smoky. *Lyophyllum fumescens* (fue MESS ens) has gills that turn a smoky color with age.

funicul (L): A small rope or cord. Species with epithets *funicularis* and *funiculosum* are typically long and thin.

furca (L): Forks. *Fuscoporia bifurcata* (bi fur CAH tah) is a dark brown-pored fungus that has setae with two prongs. In related terminology, you can use 'bifurcate' to describe something that splits in two.

furfur (L): Bran, branny, scurfy. Used in *Hymenopellis furfuracea* (fur fur AY cee uh), to denote the texture of the stipe.

fuscus (L): Brown, dark, dusky. *Scleroderma fuscum* (FUE scum) is brown.

fusus (L): Spindle. It is used in the genus *Fusarium* (few SAR ee um), one of the filamentous fungi, and a few epithets as well. In other terminology, the 'fusiform gyrus' is a long spindle-shaped structure of the temporal lobe, and *Tibia fusus* is a spindle-shaped sea sell.

G

gala / galaktos (G): Milk. *Russula galactea* (gah LAC tee uh) and numerous other species use this word element.

galerina (L): Like a cap or helmet. *Gymnopilus galerinopsis* (GAL er eye NOP sis) resembles a *Galerina*, and *Galerina* species look like helmets.

gallic (L): France. *Armillaria gallica* (GAL lic uh) was named in 1987 after the country where it was first described. Unrelated, the epithet *gallicola* refer to growth on galls.

gallopava (G): Turkey. *Verruconis gallopava* (gal lo PA vuh, formerly *Ochroconis gallopava*) is a fungal pathogen that can cause encephalitis and death in the turkey, *Meleagris gallopavo*. Although very rare, it can infect people who are immuno-compromised and people who have received organ transplants, and it can be deadly.

gambosa (L): A swelling near the hoof, club-footed. The stipe of *Calocybe gambosa* (gam BOE suh) (Commonly called 'St. George's mushroom', formerly *Tricholoma gambsum*) has a bulbous base.

gan (G): Luster/lustrous, brightness, beauty. The skin of *Ganoderma* (gan oh DER muh) is definitely bright and lustrous.

gangren (L): A rotting of flesh, gangrenous. *Calocybe gangraenosa* (gan grae NOS uh) was named after its fetid odor.

gaster (G): Stomach. The Gasteromycetes (gas TER rho my SEE tees) is a group of Basidiomycetes that have spore production within the fruitbody. Included are puffballs, earth stars, stalked puffballs, stinkhorns, and bird's nest fungi.

gausap (G): A shaggy woolen cloth. Used in *Tricholoma* and *Stereum gausapatum* (gau suh PAT um) to describe their appearance.

Gautier, Joseph was a natural historian. In 1831, Vittadini named the genus *Gautieria* (gau tee AIR ee uh) after him. I can find no more information on him, but have decided to include it because the name is so common.

gelo (L): Freeze, congeal. *Pseudohydnum gelatinosum* (gel ah tin OH sum) is like hard Jello. Related words include 'gelato' (Italian), 'gelatinous', and 'jelly'.

gemell (L): Twin. *Agaricus gemellatus* (gem el LAH tus) was named because it is very similar (is a twin with) to another species, *A. didymus*. File this under completely obscure words, but 'gemellology' is the study of twins.

gemm (L): A bud or gem. *Amanita gemmata* (gem MAH tuh) is called the 'jeweled amanita' because of veil fragments on the cap.

genu / geniculatus (L): Knee / having a protuberance like a knee. *Ophiocordyceps geniculata* (gen ick you LAH tuh) looks like a knee. In medicine and anatomy, *genu recurvatum* means 'hyperextended knee', and the 'genu' is a portion of the corpus callosum in the brain that is bent at a 90-degree angle like a knee.

geo (G): The earth, land. *Geopyxis* (gee oh PIX iss) and *Geopora* (gee oh POR uh) grow on land.

Gerard, William Ruggles (1844 - 1914) was a drug store owner from Poughkeepsie NY, who was active in natural science and botanical clubs in upstate New York and New York City. He had a working relationship with Charles Peck, and sent him many specimens, including *Lactarius gerardii*, which was originally described from Poughkeepsie. In total, there have been 40 species named after him.

geron (G): An old man. *Gerontomyces lepidotus* was found in Baltic amber, which is about 50 million years old. In the famous picture, an insect exoskeleton is found with a single mammalian hair next to the mushroom. Poinar posits that the insect had just left the mushroom, then left its exoskeleton as a piece of sap dropped and covered the scene preserving it forever. 'Gerontology' is a related word.

gibb (L): Humped, rounded, forming a swelling. Used in *Clitocybe gibba* (GIB buh) and many other species. A 'gibbous moon' is more than half full.

gigante (L): Belonging to giants, gigantic. *Calvatia* and *Aleuria gigantea* (gih GAN tee uh) (and many other species) are large.

Gilbertson, Robert Lee (1925 - 2011): Gilbertson was at the University of Arizona for much of his career, was president in the Mycological Society of America in 1980, and made significant contributions in mycology. *Laetiporus gilbertsonii* (gil bert SON ee eye) is the chicken of the woods of hardwood trees (oaks and eucalyptus) found on the coast from Mexico to Washington, and was named after Gilbertson in 2001. The original paper describing this species says "In honor of Dr. Robert L. Gilbertson … teacher and mentor to young mycologists, and valued colleague of mycologists throughout the world." (Burdsall & Banik, 2001, p. 49).

gilvus (L): Pale yellow. The margin of *Phellinus gilvus* (GIL vus) is typically the color of mustard.

glab (L): Hairless, smooth. *Strobilomyces glabriceps* (GLAB rih ceps) and *S. glabellus* are both at least somewhat hairless. 'Glabrous' is frequently used as an adjective in mycology and botany. Further, 'glabrous skin' is the opposite of skin with hair.

glac (L): Frozen, glacier. *Clitocybe glacialis* (glay see AL iss) and *Hymenogaster glacialis* are found among dwarf willows in subalpine/alpine regions.

gladius (L): Sword. *Mycena gladiocystis* (glad ee oh CYS tis) has sword-shaped cystidia. 'Gladiator' is a related word.

glaes (L): *Mycena glaesisetosa* has amber-colored pileosetae.

glandula (L). Gland. *Exidia glandulosa* (gland you LOW suh) is glandular in appearance. Further, the suffix *–osa* or *–osum* denotes fullness, so this one REALLY looks like a gland.

glans (L): An acorn. Fungi with the epithet *glandiformis* resemble an acorn. The diminutive form is 'glandula', and can refer to glands. Thus, the epithet *glanduliformis* indicates that the fruitbody resembles glands. In human anatomy, the 'glans' is indeed named after the acorn.

glar (L): Gravel. *Psathyra glareosa* (glar eh OH suh) grows on gravelly soil.

glauc (G): Silvery, gleaming, blueish-green or grey. *Phellinus glaucescens* (glau KESS ens) is variously described as iridescent or glancing (meaning gleaming or flashing), 'sometimes with a glaucous shade especially when turned in incident light' (Ryvarden and Johansen, 1980). 'Glaucose' also means pale blue green, dull or pale green, or grayish green, and 'Glaucoma' (the eye disease) is named after the resulting color/appearance of the eyes as the disease progresses.

glischr (L): Glutinous. *Limacella glischra* (GLISS chruh) is as glutinous as they come.

glob (L): Globe, ball. *Peziza globosa* (glo BO suh) is spherical. You can also use 'globose' as an adjective, if for no other reason than calling a puffball 'round' or 'spherical' lacks pizazz.

gloe (G): Glue, sticky. *Gloeophyllum* (glow ee AH fill um) has fused lamellae (they are glued together). In related terminology, glial cells in the brain, which roughly equal the number of neurons, were historically thought to be supportive and act as glue for the central nervous system.

gloss (G): Tongue. Mushrooms in the genus *Microglossum* (micro GLOSS um) look like little tongues. *Microglossum rufum* is the 'orange earth tongue'. In related terminology, the 'hypoglossal nerve' controls tongue movements.

glyc (G): Sweet. *Lactarius glyciosmus* (gly cee OHS mus) is sweet-smelling. Related words include 'glycine' and 'glucose'.

Goetz, Christel and Donald: The mushroom *Hygrophorus goetzii* (go ETZ ee eye), first described from Mt. Hood in Oregon in 1957, is named after Christel and Donald (Don). They were pioneering members of the Oregon Mycological Society. Christel was the first secretary of the Pacific Northwest Key Council, and Don Goetz was their first treasurer. Christel wrote their first key to *Phaeocollybia*, and Don wrote their key to *Amanita*. They shared the 1984 NAMA award for contributions to amateur mycology.

gomphos (G): Club or bolt. *Chroogomphus* (kroo GOM phus) species are commonly called the 'pine-spikes'.

gossypina (L): Cottony. *Psathyrella gossypina* (goss ee PEE nuh) has a cottony cap, and *Amanita gossypinoannulala* has a cottony annulus. *Gossypium* is the genus of the cotton plant.

gracilis (L): Slender. *Ophiocordyceps gracilis* (grah CILL is) is slender, as slender as any other in the genus anyway. There are a number of thin species with this same epithet.

gramin (L): Grass. *Russula graminicolor* (gram IN ih color) is called the 'grass green Russula'. This word element is used in over 750 times in fungal names! Sometimes it refers to a mushroom growing in grass, like the epithet *graminicola*.

gramm (G): A mark or a line. *Melogramma* species appear like dark marks. *Melanoleuca grammopodia* has lines on the stem, while species with the epithet *grammocephalus* have markings on the cap. In related terminology, grammar involves putting markings on paper.

grando (L): Hail. The genus *Grandinia* (gran DIN ee uh) has a hail-like granular appearance on the surface. 'Grandophobia' is the fear of hail.

graveolens (L): Strong-smelling. *Russula graveolens* (grav eh OH lens) has a pronounced odor.

gregarius (L): Belonging to a herd or flock. *Russula gregaria* (greh GARE ee uh) grows in a cluster, although I might be inclined to call a cluster of mushrooms a herd the next time I see one. Related terminology includes 'gregarious'.

griph (G): Intricate like a woven fish basket. *Grifola* (grih FOE luh) species are indeed intricate.

griseus (L): Gray. *Russula fuscogrisea* (foos co GRIS ee uh) denotes a gray color, and *grisea* or *griceus* are used over 400 times in mushroom names.

gross (L): Thick. There are many gross species, like *Amanita grossa* (gross uh), which is nicknamed the 'Tasmanian thick-stemmed Lepidella'. In other words, a 'gross' (a dozen dozen) is big. The other definition of 'gross', meaning 'disgusting', comes from the French, where it is used to describe something big or fat.

Guépin, Jean Baptiste Pierre (1779-1858) was a French medical doctor and botany professor. He was the head of the Angers botanical gardens in 1838, and published three extensive books on fungi. The genus *Guepinia* (by Fries in 1825) and *Guepiniopsis* (goo uh pin ee OP sis) were named after him.

gumm (G/L): Gummy, sticky. *Pholiota gummosa* (gum MOE suh) has a sticky pileus.

Güssow, Hans T (1879 - 1961). Güssow was a Canadian mycologist and is best known for his 1927 book 'Mushrooms and Toadstools' with W.S. Odell. The yellow variety of *Amanita muscaria, A. muscaria* var. *guessowii* (guss OW ee eye). Note that the epithet starts with the word *guess*, while the person's name is spelled with an umlaut over the *u*.

Guzman, Gaston (1932-2016): Guzman is best known for his work on psilocybin-containing mushrooms, and is responsible for discoving more than half of the world's known Psilocybe species. *Psilocybe guzmanii* (goos MAHN ee eye), *Calvatia guzmanii*, and 20 other species bear his name.

gymn (G): Naked, lightly clad. *Gymnopus* (GYM no pus) is a genus that has naked feet. No, it is not some weird kind of fetish, rather a reference to the bare stem of the mushroom. *Gymnopilus* literally means 'naked cap', and 'gymnasts' used to be naked.

gypsos (G): Gypsum, white. *Exidiopsis gypsea* (jip SEE uh) is a chalk-like color.

gyr (G): Round. *Gyromitra* (gy rho MY truh) literally means 'round hat', and *Gyroporus* has round pores. *Psathyrella gyroflexa* is named after its weak flexible stem. A related word is 'gyroscope', and even 'gyro', the Greek sandwich, is named after the meat that goes in the sandwich, which is cooked and rotated on a spit.

H

haem (G): Blood. *Mycena haematopus* (he mah TOE pus) exudes a blood-red juice when crushed.

hamadrya (G): A wood nymph, literally 'with the tree'. *Naucoria hamadryas* (ham uh DRY as) lives under trees with the wood nymph.

hamus / hamatus (L): Hook, hooked. *Coltricia hamatus* (ham AH tus) has hooked setal hyphae.

hapal (G): Soft, gentle, tender. *Hapalopilus* has a tender pileus.

hariol (L): A soothsayer. *Gymnopus hariolorum* (hare e oh LOR um) was named in 1782 (originally as *Agaricus hariolorum*) from its "fancied use by soothsayers". Another name for this was *A. sagarum*, from the Latin 'witch'. Stevenson writes "The names seem to indicate some superstitious idea attached in France to the Agaric, or some superstitious use made of it" (British Fungi, 1886).

harrya (hare EE yuh): See Thiers, Harry D.

hebe (G): Young. The genus name *Hebeloma* (he beh LOW muh) literally means 'young fringe' to describe the evanescent partial veil.

Heim, Roger Jean (1900 - 1979) was a well-published French botanist and mycologist. Although his PhD dissertation and first work was on *Inocybe*, he also worked on the Russulales, termite fungi, and hallucinogenic fungi. He traveled with Robert Gordon Wasson to Mexico to study the psilocybin mushrooms and he even cultivated *P. mexicana* and sent them to Albert Hoffmann. Heim also discovered and described a hallucinogenic species of *Boletus* from Papua New Guinea (*B. manicus*). A photograph of Heim and Wasson appeared in the famous LIFE story in June 1957. Heim is recognized with the epithet *heimii* and in the genus of Boletaceae *Heimioporus* (heim ee oh POR us).

helic (G): Coil, helix. *Helicascus* has coiled asci. There are also many species with this element in their specific epithet, like the epithet *helicospora* (he lih co SPOR uh), which denotes spiral-shaped spores. Our DNA contains the famous double helix.

helio (G): Sun. *Russula heliochroma* (heal ee oh CHRO muh) has the color and beauty of the sun.

helminth (G): A flat or round worm. The genus *Helminthosphaeria* (hel minth oh SPHAR ee uh) is an ascomycete that looks like a bunch of worms. 'Helminthoid' is an adjective used to describe something that is worm-like in form.

helos (G): A nail. The ascomycete order Helotiales (hay low tee AL es), family Helotiaceae, genus *Helotium* generally look like broad-headed nails. A popular example of the family is *Chlorociboria aeruginascens.*

helos / helodes (G): Marsh / frequenting marshes. The epithet *helodes* (hel OH dees) is used with *Cortinarius*, *Hebeloma*, and *Russula* to denote the mushroom's habitat.

helvell (L): An aromatic Italian pot-herb. The genus *Helvella* must have resembled the herb.

helvus (L): Honey yellow. *Lactarius helvus* (HEL vus) is a yellowish-honey color, while *Cortinarius helvolus* becomes pale yellow with age. This is unrelated to *Helvella* (named after a pot-herb), and fortunately there is no *Helvella helvus.*

hemi (G): Half. The genus *Hemileccinum* literally means 'half-Leccinum', and *Amanita hemibapha* means 'half-dyed'. In related words, a 'hemisphere' is half a sphere, and 'hemiplegia' is a paralysis on one side of the body.

hemorrhoid (G/F): Oozing blood. Used in several unfortunate species like *Agaricus haemorrhoidarius* (now deprecated), which bruises a blood-red, and in *Lactifluus haemorrhoeus*, which has red latex.

hepat (G): The Liver. *Fistulina hepatica* (heh PAT ih cah) is named after the liver (which produces bile). *Fistulina hepatica* is named for its shape and likeness to the liver.

hercules (From Roman mythology): *Cordyceps herculea* (her cue LEE uh) and *Cortinarius herculeus* are named after their large size. A related word is 'herculean', meaning 'requiring great force'.

hericium (L): Hedgehog. *Hericium* (hair EE see um) species are commonly called 'hedgehogs' due to their appearance.

herp (G): To creep. *Cortinarius herpeticus* (her PET ick us) (currently *C. scaurus*) has a long stipe. It is used in other fungi names and lichen as well. Related words include 'herpetology' (the study of creeping reptiles and amphibians) and 'herpes' (a retrovirus that creeps along nerves).

Hesler, Lexemuel Ray (1888 – 1977): Hesler was well published in the mushroom world. *Cordyceps hesleri* (HES ler eye), which grows on the heads of cicada nymphs in the Great Smoky Mountains, was named after him. Some of his best known publications include 'Notes on southern Appalachian fungi', 'North American species of Hygrophorus', 'North

American species of Crepidotus', 'North American species of Lactarius', and 'North American species of Gymnopilus'.

hiascens (L): Opening, splitting. The margin of *Coprinellus hiascens* splits with age. In other words, a 'hiatus' is marked by a gap in time, and a 'hiatal hernia' describes an opening in the body.

hiem (L): Winter. *Clitocybe* and *Phloeomana hiemalis* (hi eh MAL is), and *Entoloma hiemale* are all winter species. Related, 'Himalaya' is Sanskrit for 'abode of snow'.

hinnu (L): A fawn. *Cortinarius hinnuleus* (hin NU le us) is the color of a young deer.

hirn (L): A jug. *Clitopilopsis hirneola* resembles a small jug.

hirsut (L): Hairy. *Trametes hirsutum* (hir SUE tum) has hairs on the spore bearing surface. You can also use 'hirsute' as an adjective to describe any hairy mushroom (or person for that matter).

hispid (L): Spiny, shaggy, rough. *Phellinus semihispidus* (sem ee HISS pid us) has somewhat (*semi*) of an uneven texture.

von Hochstetter, Ferdinand (1829 – 1884). Von Hochstetter was a German geologist and naturalist. In 1859, he surveyed and documented the biodiversity on some of New Zealand's islands, which led to several species being named after him. Included in these are a frog, snail, and the picturesque *Entoloma hochstetteri*. Hochstetter took notes on this fungus and made a sketch of it, which he passed on to mycologist Erwin Reichardt, who named it after him in 1866. As for pronunciation, we should pronounce the name in the spirit of how Hoschstetter would have said it himself, but the best most English speakers can do is 'hoch STET er eye'.

Hodge, Kathie: Used in *Hymenoscyphus kathiae* (formerly *Pezoloma kathiae*). Richard 'Dick' Korf, professor emeritus of Biology at Cornell, said in the paper, "Named after my former student and now my successor at Cornell". Check out Kathie's awesome website and blog about mushrooms (blog.mycology.cornell.edu).

von Hohenbühel, Ludwig Samuel Joseph David Alexander Freiherr Heufler zu Rasen und Perdonegg (1817 – 1885) was an Austrian who specialized in non-seed bearing plants (cryptogams). The genus *Hohenbuehelia* (ho hen bue HEAL ee uh) was named after him.

hort (L): Garden. The epithet *hortensis* (hor TEN sis) is used with *Agaricus*, *Boletus*, *Lactarius*, *Morchella*, and many others. It is also used in the genus *Hortiboletus*. A related word is 'horticulture', meaning 'the art of gardening'.

hudnon (G): An ancient name for a fungus; a tuber or truffle. The genera *Hydnum* and *Pseudohydnum* (sue doe HID num) derive their names from this root word.

hum (L): Ground/soil. *Russula humicola* (hue MIC oh luh / hue mih COE luh) is found growing on the ground. A common related word is 'humus' (but not hummus).

humil (L): Small, dwarfish. *Tylopilus humilis* (hue MIL is) is short. *Tricholoma humilis*, although deprecated, was called the 'dwarf Tricholoma', and there are many other short species with a related epithet. Related are 'humility' and 'humble', meaning 'modest' or 'lowly'. Something that is 'humiliating' makes you feel low.

hupohana (Puebloan): Juniper. *Agaricus hupohanae* (hu poe HAN ay) is a New Mexico *Agaricus* that grows with juniper.

hyal (G): Colorless, shiny, glassy. *Hypomyces hyalinus* (high uh LINE us) is generally white. Hyaline is also a common adjective to describe colorless spores.

hydro (G): Water. *Russula hydrophila* (hi DRAH fil uh) is attracted to habitats with water like a fly to a ripe stinkhorn.

hygro (G): Wet. A 'hygrometer' measures humidity, and *Astraeus hygrometrica* (high grow MET rick us) kind of measures humidity too, as they expand on their feet when it is humid.

hymen (G): Membrane. Used in the family name Hymenochaetaceae (high men oh chee TAH cee aye). Since this words ends in *eae*, you might know that it is the taxonomic family name. This family includes the genera *Coltricia*, *Inonotus*, and *Phellinus* to name a few of the more common ones. The first part of the family name means 'membrane'. Related, the Hymenoptera (bees, ants, and wasps) are literally the 'membrane-wings'.

hyph (G): Thread. *Hypholoma* gives a reference to the thread-like veil remnants on the margin of the cap.

hypnon (G): Tree-moss, including the genus *Hypnum*. *Hypnorum* is used in nearly 100 species that are associated with moss, including *Galerina* and *Clitocybe hypnorum* (hip NOR um).

hypo (G): Under. *Scleroderma hypogaeum* (high poe GAE um) typically grows underground. 'Hypogeous' is frequently used as an adjective to describe an underground fruitbody, like a truffle or a *Rhizopogon*.

hyps (G): High. *Hypsizygus* (hip sih ZYE gus) species are often found high up in the trees.

hysginon (G): A scarlet or crimson dye, dark reddish pink. *Russula hysgina* (hiss GHI nuh) is scarlet-red. In other terminology, *Dario hysginon* is a scarlet-colored zebrafish (at least the males are scarlet, anyway).

hyster (G): The womb. The truffle *Hysterangium* (his ter AN gee um) appears to be like a womb. Related words include 'hysterectomy', and 'hysterical'.

hystrix (G): Porcupine. *Hericium hystrix* (HIS trix) has a split personality: the name literally means 'hedgehog porcupine mushroom'.

I

ianth (G): Violet-colored, also see 'janth'. Dozens of species starting with this word are various shades of violet, like *Leucocoprinus ianthinus* (i an THIGH nus).

icter (G): Jaundice, the yellowish color associated with jaundice. *Dermocybe icterinoides* (ick ter in OY dees) and *Psora icterica* exhibit various shades of light yellow green. 'Icterus' is the medical term for jaundice.

idius / idia / idium (G): A common diminutive suffix. For example, *Otidia* (oh TID ee uh) means 'little ear'.

igneus (L): Fire. *Cantharellus ignicolor* (IG nih color) (formerly called *Craterellus ignicolor*) is similar to *Craterellus tubaeformis*, but has more yellow and orange in the cap. That is, it has more flame-related colors. In related terminology, 'igneous rock' forms from the solidification of magma or lava.

ile/ileum/ilium (L): Entrails. The name *Ileodictyon* (il lay oh DIC tee on) suggests that the net (dicty) looks like entrails. The specific epithet *ileiformis* more directly suggests a fruitbody looks like intestines. Indeed, the 'ileum' is a portion of the small intestine.

illuden (L): To deceive. *Omphalotus illudens* (il LUE dens) is toxic and is mistaken for eatable chanterelles.

imbric (L): To cover with tiles or scales. *Sarcodon imbricatus* (im brih CAY tus) (common name: Hawk's wings) is covered with scales.

Imler, Louis Philip Mathieu (1900 - 1993) was a Belgian mycologist and founder of the Anterp Mycological Circle. He named *Xerocomus porosporus* in 1958, and the genus of Boletaceae *Imleria* (im LER ee uh) was named in his honor in 2014.

imperialis (L): Noble, imperial. *Gymnopilus imperialis* (im pier ee AL is) is a stately fungus – as are the other 52 hits for this epithet on Indexfungorum.org.

impolit (L): Unpolished, rough. *Ganoderma impolitum* (im poe LITE um) has a rough surface.

importun (L): Assertive, inconsiderate, troublesome, importune. *Morchella importuna* (im por TOO nuh), the so-called 'landscape morel', was named after its "inconsiderate" encroachments on gardeners and homeowners. I'm sure I'm not the only one who would happily come to the rescue and remove these heedless interlopers from property if needed.

impudicus (L): Shameless, immodest. *Phallus impudicus* (im PUE dic us) literally translates to 'the shameless penis'.

in (L): Without. A common prefix used in many mushroom names. For example, it is used in the species *Cortinarius* and *Myochromella inolens* (in OH lens) to indicate that they lack an odor. This is the same prefix used in 'inflammable', 'inexpensive', and 'insane'. Note that same prefix also means 'in', or 'within' in other cases.

incan (L): Hoary (grayish white). *Entoloma incanus* (in CAY nus) is grayish white at maturity.

incarnatus (L). Made flesh, made red, reddish. *Russula incarnaticeps* (in car nat ih ceps) has a reddish cap. In other terminology, the word 'incarnate' means 'embodied in flesh', yet the third definition of the word just means 'flesh-colored or crimson'.

incert (L): Uncertain. Peck described and named *Hypholoma incertum* (in CER tum), but was unsure whether it was distinct from *H. candolleanum.* Although they are both now *Psathyrella* species, they are distinct from each other. About half of the 70 species that have had this uncertain designation have been renamed in no uncertain terms.

incis (L): Cut. The cap of *Cortinarius incisus* (in CYE sus) tears into fibrils in dry weather. A related word is 'incisor', the teeth you'd use to rip into this species if it was edible.

inconstans (L): Fickle, unsteady, inconstant. *Russula inconstans* (in CON stans) can present itself in various forms. I'm guessing that Murrill noted some differences between specimens in his collection.

incrassatus (L): Thickened. *Auricularia incrassata* (in cras SAT uh) is thick. The various forms of the word include *incrassata*, *incrassatum*, and *incrassatus* get 129 hits on indexfungorum.org, so this is a fairly common descriptor.

indecorat (L): Ugly. *Tremella*, *Puccinia*, and *Russula indecorata* (in deh cor AH tuh) are ugly – or at least someone thought that when the species were named.

indigo (L): A dark blue dye. *Lactarius indigo* is the 'indigo milk cap', and it is deep blue. The latex is also blue, but turns green over time.

indusiatus (L): Undergarment-wearing. The species name *Phallus indusiatus* (in DUSE ee AH tus) literally means 'the penis wearing an undergarment'. However, the 'undergarment' part of the name actually refers to the delicate lace-like skirt draping over the mushroom.

infest (L): Infest. *Phytophthora infestans* (in FES tans) attacks potatoes.

infract (L): Broken, bent. The margin of *Cortinarius infractus* (in FRAC tus) is broken and bent, at least in mature specimens. In related terms, if you've gotten a traffic infraction, you probably broke a law.

inful (L): A band, a woolen band that is part of a religious head-dress. *Gyromitra infula* (in FEW luh) is commonly called the 'elfin saddle' or the 'hooded false morel'. 'Infula' has many applications, including in a woolen head-dress worn by Roman priestesses, the vestal virgins, to highlight their vows of chastity. The wool coiled around the head a number of times and draped over the shoulders. There are also two infulas (wool bands) hanging down from a bishop's mitre. Either way, the draping sides of *G. infula* is reminiscent of the woolen bands.

infundibul (L): A funnel. *Clitocybe infundibuliformis* (in fun dib you lih FOR mis) has funnel-shaped mushroom caps. There are 108 records with *infundibuliform* in the name, including *Clitocbye*, *Cantharellus*, and *Polyporus* species, just to name a few.

infusc (L): Darkened. *Amanita infusca* (in FUES cuh) is called the 'dark slender Caesar', and some form of 'infusc' is used in over 50 other darkish species.

ingen (L): Great, remarkable, large. *Russula ingens* (IN gens) and 14 other large or otherwise remarkable species share this epithet.

ingratus (L): Unpleasant. Russula section Ingratae (in GRA tay) (Singer, 1986) typically have prominent, and sometimes unpleasant odors, like the foetid Russula.

innixus (L): Leaning, reclining. *Aureoboletus innixus* (in NIX us) (formerly *Boletus innixus*) has a tendency to lean over when they grow, especially when they grow in clusters.

ino (G): Fiber/fibrous, muscle. The genus name *Inonotus* (eye NON oh tus) means 'fibrous ear'. The genus *Inocybe*, comprises the 'fiber-heads'.

inornat (L): Unadorned, undecorated, not beautiful. *Clitocybe inornata* (in or NAT uh) (now *Atractosporocybe inornata*) is, for all intents and purposes, plain.

inquilin (L): To inhabit, a tenant. Forty-two species have been given this name to merely describe that they live on something, like *Deconica inquilina* (in quil EE nuh). The Spanish word 'inquilino' means 'tenant'.

inquinat (L): Polluted. *Podospora inquinata* (in quin AH tuh) grows in polluted habitats.

insign (L): Unique, decorative. *Leccinum* and *Calostoma insigne* (in SIG nee) are unique indeed. In related words, an 'insignia' is a distinguishing mark, emblem, or badge of honor.

insipid (L): Tasteless, insipid. *Hygrocybe insipida* (in SIP id uh) is 'absolutely tasteless'.

integ (L): Entire, perfect, spotless. *Russula integra* (in TEG ruh) is called 'the entire russula', and one source called it 'perfect in form'. *Delicatula* (formerly *Mycena*) *integrella* is entirely small. An 'integer' is a whole number, and someone with 'integrity' is whole. 'Integrate' and 'integral' are related too.

inter (L): Between. 'Inter' is used in over a thousand fungal species (and hundreds of common words). For example, *Tricholoma intermedium* (in ter MED ee um) is somewhere between *Tricholoma equestre* and *sejunctum*.

interced (L): A combination of 'inter' (between) and 'cede' (go), together 'to come between'. The 50 plus species with the epithet *intercedens* (in ter SEE dens) are intermediary between two other species. 'Intercede' means 'to mediate'.

interrupt (L): Broken. *Mycena interrupta* is so picturesque that it is often posted with the line 'is this mushroom real'? It is real, but what about the derivation of the specific epithet? Saccardo concluded his description of the mushroom in his 1887 publication with 'pilei carne gelatinoso-carnosa, descendente interruptis'. This last clause indicates that the cap appears 'broken down', perhaps upon drying or with age. In related terms, if you interrupt someone, you break into their conversation.

inus/-ina/-inum (L): A suffix added to nouns to denote a belonging to or resemblance. It is used in many genera including *Coprinus*, which literally means 'belonging to dung', not 'dung loving' as I frequently hear (that would be 'coprophilic').

involut (L): Wrapped up, involute. *Paxillus involutus* (in voe LOO tus) has a prominent involute margin (one that is rolled inwards, or incurved).

io (G): Violet. *Gymnopus iocephalus* (eye oh ceph AL us) has a violet cap.

iocular (L): Funny. Many mushrooms look funny, some make you feel funny, and some names are funny. Like the 2-inch stinkhorn named after someone who really wanted a mushroom named after themselves, or when Burgeff created the epithet *blakesleeanus* as a personal insult to Blakesley. Then there are names where someone tried to be funny but really wasn't. Like naming a mushroom species after a wasp because the collector got stung when they collected it, or using the epithet *ridicula* because you acknowledge it is ridiculous you are publishing a new species based on a single collection. Only one species I have found is actually named for looking funny, though: *Coprinopsis iocularis* is literally 'the funny Coprinopsis'. It was named after the 'funny shape of the spores'.

irina (L): The iris (the plant). *Lepista irina* has a flowery odor, like an iris.

iris (G): The rainbow, or the plant, which is itself named after the rainbow. *Melanoceuca* and *Cortinarius iris* are multi-colored, while *Puccinia iridis* is a rust on the iris plant and *Cercospora iridicola* grows on iris leaves. Related words include the 'iris' (of the eye), 'iridescent', and 'Iris', the messenger of the gods.

irpex (L): A harrow. The genus *Irpex* (IR pex) has rows of teeth like a harrow.

irresolut (L): Doubtful. The placement of *Gymnopus irresolutus* (ir res oh LUT us) in *Gymnopus* is irresolute, or uncertain.

irrig (L): Related to water. *Gliophorus irrigatus* (ir rih GAY tus) has a viscid wet-looking cap. Naturally, a related word is 'irrigation'.

is / iso (G): Equal. *Aleuria isochroa* (ice oh CROW uh) means 'of uniform color'. Related, an 'isosceles triangle' has two sides of equal length.

isabelline (E): Pale cream-brown, color of sole-leather. There are over 125 records of this as a species name, including *Peziza isabellina* (is a bell EE nuh), *P. subisabellina*, and *Cortinarius isabellinus*. How Isabell became associated with this color is unclear, but it seems it was used to describe the color of her clothes after wearing them for a protracted period.

ischnos (G): Thin. *Scutellinia ischnotricha* has thin hairs and *Hebeloma ischnostylum* (ish no STY lum) has a thin stipe.

issimus (L): a suffix denoting very much, most. *Phallus tenuissimus* (TEN you ISS im us) means that it is the 'most slender'. *Phallus tenuissimus* measures in at a whopping 2 inches long, perhaps the true micro-phallus.

ite (G): Denoting a member of a group or part of something, also to denote the fossil nature of something. This suffix is widespread in the sciences and taxonomy and naming in general, including in mushroom names. *Coprinites* (co prin ITE ees) *dominicana* was a Coprinaceae-like mushroom found in Dominican amber and is about 40 million years old.

iuvenis (L): Young. *Cantharellus iuventateviridis* (you ven tate VIR ih dis) is greenish when young. I've also seen 'iuventate' used as an adjective to refer to a young specimen. In one write-up, I saw "iuventate creama", meaning "cream-colored when young".

J

Jackson, Henry Alexander Carmichael (1877 – 1961): Jackson is best known for the publication 'Mr. Jackson's Mushrooms' that was posthumously published in 1979. The cover of this book features Jackson's painting of the mushroom we now know as *Amanita jacksonii* (jack SONE ee eye), which was named after him in 1984.

jafnea: Richard Korf named the genus *Jafnea* (jaf KNEE uh) after J.A.F. Nannfeldt.

Jahn, Hermann Theodor (1911 - 1987) was a German mycologist and ornithologist who published over 160 works during his career. *Jahnoporus* was named after him in 1980, and many specific epithets bear his name as well.

janth (G): Violet-colored, and an alternate spelling of 'ianth'. Violet colored species include *Penicillium janthogenum* (jan tho GEN um), *Gymnopilus janthinosarx*, and *Ombrophila janthina*. *Janthinobacterium* is a genus of soil-dwelling purple bacteria, hypothesized to infect some fungi.

jasonis: A reference to Jason's golden fleece of Greek mythology. The cap color of *Cystoderma jasonis* (jay SONE iss) is reminiscent of the famous golden fleece.

jub (L): Crested, maned. *Entoloma jubatum* (ju BAT um) is the 'crested Entoloma'. 'Jubate' is an adjective meaning 'having a mane'.

judae: This is named after Judas Iscariot. The epithet of *Auricularia auricula-judae* (au RIC you luh - JEW day) is a reference to the elder tree upon which

Judas Iscariot hanged himself. As the legend goes, the ear-like mushrooms that come from the trees are vestiges of his spirit. As such, this should be called the 'Wood-ear jelly fungus', The common name 'Jew-ear', although common in 1950's era mushroom books, is incorrect, derogatory, and should not be used.

juglan (L): A walnut tree or walnut. *Microstroma juglandis* grows on walnut leaves. In total, around 250 species have had this or a related epithet, named after the genus name of the walnut tree, *Juglans*.

julius: The month of July, named after Julius Caesar. *Agaricus julius* (JU lee us) fruits in July, and *Agaricus augustus* (named after Caesar Augustus) fruits in August. The true prince is *Agaricus augustus*, so what about *A. julius*? Kerrigan has floated the name "the Emperor formerly known as Prince".

junc (L): Referring to the Juncaceae, the rush family. Epithets *juncina* (jun CEE nuh) and *juncicola* grow in association with the rushes.

Junius, Hadrianus (1511-1575): *Phallus hadriani* (had ree ANN ee) was named after Hadrianus Junius. Hadrianus was a Dutch scholar who wrote a pamphlet in 1564 on the stinkhorn that now bears his name. Although factually incorrect, this was the first (know) European monograph on a fungus.

juno (A Roman Goddess): The wife of Jupiter, often used to highlight beauty, being notable, or indicating youthfulness or rejuvenation. For *Gymnopilus junonius* (ju NO nee us), the subject of the epithet refers to the physical fruitbody, not the hallucinogenic properties the mushroom bestows. The deprecated name for this species, *Gymnopilus spectabilis*, also means 'remarkable'.

K

Konrad, Paul (1877 – 1948): Konrad was a French mycologist. He is known for a few books, including 'Icones Selectae Fungorum', or 'Pictures of Selected Fungi'. *Craterellus konradii* (con RAD ee eye), and a few other species, are named after him.

Korf, Richard "Dick" Paul (1925 - 2016) was an American mycologist and professor emeritus at Cornell University in New York. He co-founded the journal *Mycotaxon* and published numerous papers. *Gyromitra korfii* was named after him in 1973, and other species are named after him as well. He published one species himself: *Arachnopeziza fitzpatrickii.* Kathy Hodge was one of his former students and was his successor at Cornell.

Kretzschmar, Eduard: Curtis Gates Lloyd wrote in 1919 (in The Large Pyrenomycetes) that Kretzschmar was a "traveler and botanist", but "never wrote a line on mycology in his life, nor collected a fungus as far as I ever noted, and has no more claim to be 'honored' in connection with mycology than he has in aviation". This was nothing personal against Kretzschmar, Lloyd just had a hardline approach against naming species after people. Regardless of what Lloyd spewed out in his self-published works, *Kretzschmaria* (kretzsch mar EE uh) was named after the first person who found it.

Krieger, Louis Charles Christopher (1873 - 1940) was an American mycologist and author. He was also a prolific botanical illustrator who also did portraits to pay the bills. Krieger illustrated hundreds of fungi, and several of his pieces graced an issue of National Geographic in 1920. In the late 1910's, he was charged with organizing Howard Kelly's mycology library, which was no small feat because it was the largest one known. Upon Kelly's death, the library was donated to the University of Michigan, where it was called the "L.C.C. Krieger Mycological Library". *Agaricus kriegeri* (KREE ger eye) is one of a number of other species named after him.

Kuehner, Calvin C (1922 – 2011) was an American mycologist who was born in Ohio and went on to earn his B.S., M.S., and Ph.D. from the Department of Botany and Plant pathology at Ohio State University in Columbus. He published many articles and book chapters, and the genus *Kuehneromyces* was named after him. This Kuehner was confused with the French mycologist Robert Kühner at least once on the internet.

Kühner, Robert (1903 – 1996) was a French mycologist who published 160 articles and books on fungi, including several monographs and works on fungi of specialized habitats including subalpine forests. He described and named *Marasmius epidryas.* A note that *Kuehneromyces* was named after the American mycologist Calvin Kuehner.

L

lacc (Italian/French): Varnish. *Ganoderma cupreolaccatum* (coo prey oh lah CAH tum) looks like it has a copper-colored varnish.

lacer (L): Torn, tangled, mangled. *Clitocybula, Calvatia, Trametes, Puccinia*, and *Pyrenopeziza lacerata* (lah cer AH tuh) each appear torn. Related words include 'lacerated' and 'lacerations'.

lachn (G): Woolly hair. *Ophiocordyceps lachnopoda* (LACH know POE dah) has a woolly stem. *Lachnellula* also uses this word element.

lacrima (L): Tears. *Mycena lacrimans* (LAH crih mans) is a bioluminescent fungus that exudes a substance from the cystidia. In related terminology, our lacrimal gland produces tears and our lacrimal duct drains them into our sinus cavity.

lacun (L): Ditch, pit. *Helvella lacunosa* (lah coo NO suh) has a stem that is full of pits. The English word 'lacuna' signifies that something that is missing, and it can represent anything from a few years of history to a missing portion of a manuscript. In medicine, a 'lacunar stroke' is a blockage of a deep penetrating artery that leaves an empty space in the brain where the tissue supplied by artery dies.

lacust (L): Of lakes. The epithet *lacustris* is used in about 75 fungal species and throughout biological nomenclature that relate to the habitat. 'Lacustrine plains', 'Lacustrine deposits', and 'Lacustrine ecosystems' are all related phrases.

laet (L): Gay, pleasing, abundant, bright. Although I am gay with joy, overcome, and pleased when finding *Laetiporus* (lay et ih POR us) *sulfureus*, the species is named after its bright pores.

laev (L): Smooth. Used in *Crucibulum laeve* (LAY vey), the 'bird's nest fungus'. *Scleroderma laeve* is from the western United States and is smooth, at least at first.

lagena (L and G): A flask. Used in *Inocybe lageniformis* (lah gen ih FOR mis) and *Geastrum lageniforme* to indicate something about them are flask-shaped.

lago (G): A hare. *Coprinopsis lagopus* (LAH go pus) is called 'the hare's foot Coprinus', because the young fruitbody looks like a rabbit's foot - as much as one can, anyway.

lamin (L): Thin layer. *Peniophora laminata* is thin. *Sparassis laminosa* (lam in OH suh) is made up of thin plates. Related words include 'laminate' (to put a thin covering around something) and 'laminate flooring'. Unrelated is the epithet *lamingtonensis*, which indicates the species is named after Mount Lamington in Papua New Guinea.

lampros (G): Bright, clear, shining, beautiful. *Russula lamprocystidiata* (LAM pro sis tid ee AH tuh) has bright/clear cystidia.

lana (L): Wool. *Cortinarius laniger* (LAN ih ger) has a wooly pileus and *Coprinus laniger* fruits from a wooly mass of mycelium. Further, the bases of

Gymnopus and *Mycena lanipes* are wooly. *Floccularia albolanaripes* has a white wooly foot.

lancea (L): Lance, spear. *Psilocybe semilanceata* (sem ee lan cee AH tuh) looks somewhat like a spear.

languidus (L): Faint, dull, slow. *Russula languida* (LANG guid uh) is either underwhelming or has a dull color. In related terminology, 'languid' means 'slow, listless, or lacking enthusiasm'.

lanug (L): Wooly. *Inocybe lanuginosa* (lan you gin OH suh) is flocculose. 'Lanugo' is a wooly hair that can be present on a newborn, or it can develop on someone with anorexia.

laracin / larix (L): Larch tree. *Russula laracinoaffinis* (lar ah SEE no af FIN is) grows near Larch trees. It is also used in *Russula laricina* (lare ih SEE nuh). In related terminology, the word 'affinity' means 'a natural liking or relationship other than by blood'.

Lasch, Wilhelm Gottfried. (1787 - 1863). Lasch was a German pharmacist, botanist, and mycologist. Little is known about him as a person, but his name lives on in many mushroom names. His name is used in the genera *Laschia* and *Favolaschia* (Honeycomb + Laschia), as well as in the species *Puccinia* and *Typhula laschii*. *Pholiota gummosa* was originally described by him.

lasi (G): Hairy. Members of the genus *Lasiosphaeria* (laz ee oh SVER ee uh) have hairy perithecia.

lat (L): Broad. *Helvella latispora* (lah tih SPOR uh) has fairly plump spores. Lines of latitude on the globe are the broad ones that go around it.

later (L): Brick, tile. *Cantharellus lateritius* (lah TER ih TEE us) is commonly called the 'Smooth Chanterelle', and the species name means 'bricklike'. It is named after the fairly smooth underside, or hymenium. The more commonly known species *Hypholoma sublateritium*, 'the Brick-top' also has this word element in it, but here, it refers to the brick color of the cap (Metzler & Metzler, 1992).

lateris (L): Side, lateral. *Laterispora* (lah ter ih SPOR uh) has spores arising from the side.

latern (L): Lantern. *Laternea* species look like lanterns.

lauroceras (L): The cherry laurel of Asia and Europe, *Prunus laurocerasus*. Species like *Lentinellus laurocerasi* grow on the trunk of this tree.

lavendula (L): Lavender. *Amanita lavendula* (la VEN doo luh) stains lavender, while *Tricholoma lavendulophyllum* has pretty lavender gills.

lax (L): Loose. *Corticium laxum* (LAX um) comes off its substrate easily. 'Laxative' is a related word.

lazul (L): Azure blue, aquamarine, sky blue. *Mycena lazulina*, known as 'tiny blue lights', and *Cortinarius lazulinus* are a couple of species that represent these beautiful shades of blue.

Lea, Thomas Gibson (1785-1844): Thomas Lea was an amateur mushroomer and found some pretty cool mushrooms. His collections ended up being in the hands of Miles Joseph Berkeley, and Berkeley named *Mycena leaiana* after him. Unfortunately, Lea died a year before the mushroom was named after him.

leccino (Italian): Fungus. The genus *Leccinum* is just from an Italian word for fungus.

Lecomte, Paul Henri (1856 - 1934). Lecomte was an accomplished French botanist who authored 15 books during his career. *Panus lecomtei* is named after him. This mushroom is found from Mexico to Wisconsin, and in the eastern United States, and is probably more common than suspected.

Le Gal, Marcelle Louise Fernande (1895-1979) was a French mycologist. The genus *Galiella* (gal ee EL uh / gal ee AY uh) was named after her by Korf and Nannfeldt, and *Rubroboletus legaliae* is named after her as well. Her first interest in mycology was fueled by mycophagy – to identify species her husband wanted to consume. However, she soon become engrossed in the field and became a world authority on discomycetes. Notably, she was the first woman president of the French Mycological Society.

lent (L): Pliable, tough. Used in the genus *Lentinus* (len TIE nuss) and related genera (*Neolentinus*, *Lentinellus*, etc…).

Lenz, Harald Othmar (1798 - 1870) was a German zoologist, botanist, and mycologist who published several fairly extensive books. *Lenzites* (Lenz EYE tees) was named after him.

leo (G): Lion. *Hyphoderma leonium* (lay OH knee um) is colored like the pelt of a lion. 'Leo' is the lion constellation, and the 'Leonids', an annual meteor showers, originates within the constellation.

leot (G): Smooth. *Leotia* (lay OH shuh) *lubrica* and *L. viscosa* are as smooth as a mushroom can be.

lep (G): Scale. *Lepiota* (lep ee OH tuh) literally means 'scaly ear', and the word element is used in many other specific epithets as well.

lepidus (L): Pretty, pleasing. *Russula lepida* (leh pid uh) is a pretty fungus, and *Russula lepidicolor* has pretty colors.

lepist (L): Goblet. *Lepista* (leh PEE stuh) has some members that look like a goblet when they are mature.

lepor (L): A hare. *Otidia leporina* (lep or EYE nuh) looks like rabbit ears. Together, the full name literally means 'little rabbit ears'. Rabbits are in the taxonomic family Leporidae.

lept (G): Thin, small. *Polyporus leptocephalis* (LEP toe ceph AL iss) (a former *Coltricia* species) has a thin cephalis (head, or cap). A related word is 'Leptin', a hormone that causes one to be thin.

Le Rat, Auguste-Joseph (1872 – 1910) was a French botanist who also collected mosses, shells, insects, birds, and minerals. He was in good health in June of 1910, but contracted a chest condition a few weeks later and died while still in his 30's. The genus *Leratiomyces* (lay rat ee oh MY seas) was named after him.

leucos (G): White. The genus *Leucoagaricus* (LOU co a GARE ih kiss) is a common genus of mushrooms that have white caps, white gills, and white stems. They also have an annulus and are often misidentified Amanitas. However, *Leucoagaricus* do not have a volva, among other things. In related words you might know, 'Leukemia' is a disease that results in a high number of abnormal white blood cells.

levis (L): Light, small. *Amanita levistriata* has faint striations, particularly when young.

liber (L): To set free, liberated. *Morchella semilibera* (sem ee lie BER uh) is called the 'half-free morel' because the lower half of the stem is free (liberated) from the cap.

lichen (G): Lichen, from 'leikhen', meaning 'what eats around itself', possibly from 'leichein', meaning 'to lick'. Lichen are typically composed of both a fungus and an alga. There are also over 300 fungal species that utilize 'lichen' in the epithet. Species including *lichenicola* grow with lichen, and those with *licheniforme* look like lichen.

lign / lignum (L): Wood. *Buchwaldoboletus lignicola* (= *Boletus lignicola*) (lig NIC oh luh) is found on wood. Also, *Clavariadelphus lignicola* grows on wood or spruce needles, and *Ganoderma lignosum* (lig NO sum), which grows on wood. You may be familiar with the word 'lignin', a compound found in the cell walls of plants that makes them woody.

lilacina (L): Lilac, lilac-colored. *Russula citrinolilacina* (sih TREE no lie luh CEE nuh) has shades of yellow and lilac. *Leucocoprinus lilacinogranulosus* is a lilac-colored relative of the uber-common *L. birnbaumii* (which is one of the top 10 most frequently posted mushrooms in online identification groups).

lilliput: After the land of Lilliput in Gulliver's Travels. *Psathyrella lilliputana* is little enough for the Lilliputians of Lilliput.

limac (L): A slug. Used in a bunch of slimy mushrooms, like *Lactarius limacinus* (lie muh SEE nus) and *Hebeloma limacinum*. Note that many of the species with this epithet are only viscous when they are moist. In case you didn't suspect it, *Limax* is a genus name of slugs.

limbat (L): An edge, border. *Russula viridirubrolimbata* (vir id ih roo bro lim BAH tuh) must have a greenish-red border. A related word is 'limbo', either an uncertain period of time, or the place at the edge of hell where unbaptized people go (or something like that).

limon (French): Lemon-colored. *Cortinarius* and *Tricholoma limonius* (lie MOE knee us) are lemon-colored, while *Hymenopellis* and *Mycena limonispora* have lemon-colored spores.

lingzhi (Chinese): *Ling* = spiritual, *zhi* = plant/mushroom. Used in *Ganoderma lingzhi* (ling zee), which was described from China. *Lingzhi* is also the common name for several *Ganoderma* species.

liquirtia (L): Licorice. *Gymnopilus liquiritiae* apparently tastes like licorice.

lith (G): Stone. *Pisolithus* (PIE so lith us) literally means 'pea stones', due to the stone-like pea-shaped peridioles inside. 'Lithography' and 'lithograms' originally used stones as a printing surface.

litor / littor (L): Sea shore. *Inocybe, Agaricus,* and *Leucoagaricus littoralis* (lit tor AL is) grow on the coast.

livid (L): Lead-colored, bluish-gray. *Lycoperdon lividum* (LIV id um) is named after its color.

lix (L): Ashes, lye. *Clitocybe lixivia* (lix IV ee uh) is ash-colored.

Lloyd, Curtis Gates (1859 – 1926): *Tulostoma lloydii* (LLOYD ee eye) was named after him. Lloyd was most interested in the Gasteromycetes and published extensively. He had his own journal, *Mycological Notes*, where he described numerous species and bashed fellow mycologists he disagreed with.

lobos (G): A lobe. *Ganoderma lobatum* (low BAH tum) has lobes.

lom / loma (G): Fringe, border. *Ganoderma megaloma* (meg ah LOME uh) is found in the eastern part of the United States and differs from *G. applanatum* by having a wide (thus the *mega* part of the species name) sterile border around the edge.

loricat (L): Armored. *Lyophyllum loricatum* (lor ih CAT um) has a thick cartilaginous cuticle and is called the 'armored Lyophyllum'. A 'lorica' refers to several pieces of body armor.

lubric (L): Smooth, slippery. Both parts of the species name of *Leotia lubrica* (LOO brih cuh) mean smooth. Indeed, the 'jelly baby' is smooth. Related are 'lubricant' and 'lubritorium'.

luc (G): Light. *Lactarius luculentus* (loo cue LEN tus) is 'full of light'. Related words include 'lux', 'luculent', and even 'Lucifer'.

lucidus (L): Bright, shining. The epithet of *Ganoderma lucidum* (LOO sid um) literally means 'shiny bright skinned'.

ludovic (L): The Latinized form of Louisiana, itself named after King Louis the XIV of France. There are about 40 fungal species with this epithet that come from this part of the United States, including *Agrocybe* and *Gomphus ludoviciana* (loo doe vic ee AN uh), as well as *Pluteus ludovicianus.*

lugen (L): To mourn. *Hebeloma lugens* (LOO gens) has a gloomy somber color.

luridus (L): Pale yellow. *Russula lurida* (LURE id uh) has a pale-yellow pileus.

lustratus (L): Purified, white. *Cortinarius lustratus* (lus TRAT us) was named because of its lack of color. This word is related to 'luster', from the Latin *lustrare*, 'to illuminate or brighten'.

luteo / luteus (L): Yellow, pale yellow, saffron-yellow. *Russula luteobasis* (loo tee oh BASE is) is yellow at the base of the stipe. *Phallus luteus* (LUE tee us) a species with a beautiful yellow skirt. *Gymnopilus luteus* and *G. luteofolius* are also yellow.

luteol (L): Pale yellow. *Luteolus* is the diminutive form of 'luteus' and indicates a paler yellow color. It is also used in *Inocybe luteola* (loo tay OH luh), *Entoloma luteolamellatum* (pale yellow gills), and *Galerina luteolosperma* (pale yellow spores).

lycos (G): Wolf. Used in the genus *Lycoperdon* (LIKE oh PER don). The genus name of several common puffballs, derives the first part of its name from the wolf. You can also call someone 'lycanthropic' if you want to point out their similarity to a wolf. Someone interviewing for a job at my previous place of employment referred to my boss as 'lycanthropic'. Although she was trying to impress us with her vast vocabulary, she didn't end up getting the job.

lyo (G): Loose, free. *Lyophyllum* literally means 'loose gills', although I am unclear of the application in this particular name.

Lyon, H. L. The truffle *Tuber lyonii* (= *Tyber texense*) was discovered near Minneapolis, Minnesota on March 11th, 1903 by Lyon. It was discovered under basswood trees, but this is the same fungus found in southern pecan orchards that is called the 'Pecan Truffle'.

lys (G): Loosen. *Lysurus* has loosened lobes sticking up into the air. Related words include 'lysis', 'lysosome', and 'lysine'.

M

macula (L): Spot. *Russula luteomaculata* (loo tay oh mac you LAH tuh) has yellow spots. Thinking of yellow spots, the 'macula lutea' is a yellow spot at the focal point of your retina.

magn (L): Great, large. *Tricholoma magnivelare* (mag ni veh LARE ey) has a great big veil. *Inonotus magnus* (MAG nus), *Mycena magna*, and over 650 other species names listed on indexfungorum.org are either large or have a particularly large feature. The Latin phrase '*magna cum laude*' means 'with high honors'. Related English words include 'magnificent', 'magnify', and 'magnifying glass'.

magnolia (L): A genus of plants, itself named after the French botanist Pierre Magnol (1638 – 1715). *Xylaria magnoliae* (mag NO lee aye) grows on magnolia fruits.

mai (Japanese): Dance. The common name of *Grifola frondosa* is 'maitake' (my TOCK ee), which literally means 'dancing mushroom'. This is said to refer to the frilled or skirt-like appearance of the individual fronds, and the movement of them as the mushroom is collected. Other places on the internet say it is named because of the dance collectors do when then they find this mushroom.

majal (L): Pertaining to May. Many species originally found in May have the epithets *majale* or *majalis*, including *Entoloma majale* and *Galerina majalis* (MADGE uh lis).

malicor (L): The covering or rind of the pomegranate. The pileus of *Cortinarius malicorius* (mal ih COR ee us) is colored like a pomegranate. 'Malachor' appears to be unrelated.

malus (L): Bad. *Melanoleuca malodora* (mal oh DOR uh) has a bad odor, and *Agaricus malangelus*, which was originally described from Angel Fire, New Mexico, and is called the 'bad angel'. Related words include 'malaria', 'malfunction', 'malady', 'malice', and 'malpractice' to name a few.

malve (L): Mallow, the color of the mallow plant, mauve, pale purple. The word is used as an adjective and in a few epithets to denote this particular color, including *Cortinarius malvaceus* (mal VAY cee us).

Mao, Lan (1397 – 1476) was a Chinese botanist of the Ming dynasty. Mao was interested in Boletes and also published a book on local folk medicine. *Lanmaoa* (lan MAO uh) was erected in 2015 along with several other genera of Boletaceae. There are four species in the United States, *L. pseudosensibilis*, *palidoroseus*, *carminipes*, and *borealis*. All have shallow pores, a blue-staining yellow pore surface, and a reddish cap.

mapp (L): A table napkin, cloth, white cloth. *Amanita mappa* (MAP puh) has a napkin-like volva. The species is now called *A. citrina*, but the specific epithet is used in other species as well. Indeed, the word 'map' comes from this origin, as they were originally made on white cloth.

marasm (G): Wasting, withering. Members of the genus *Marasmius* (mar AS me us) readily dry out but come back to life with moisture. If you haven't dried a specimen out and then reconstituted it in a bowl of water, you're definitely missing out. In related words, 'marasmus' is the medical word to describe someone who is severely undernourished and underweight.

marchantiae (L): Named after the genus of Liverwort, *Marchantia*, which was itself named after the botanist Nicholas Marchant. *Loreleia marchantiae* (mar CHAN tee ay) is a small species that grows with the liverwort.

margar (G): A pearl. *Inocybe margaritispora* has beaded spores.

marginat (L): Border, edge, margin. *Galerina marginata* (mar gin AH tuh) often has remnants of the veil on the margin of the pileus.

marmor (L): Marble. *Amanita marmorata* (mar mor AH tuh) has a marbled pileus. This is used in over 100 other fungal species too. The genus of the groundhog, *Marmota*, is unrelated, as that comes from an Indian word meaning 'digging'.

mars (L): Mars. *Lepiota martialis* literally means 'the Lepiota from Mars', but perhaps 'the Mars-like Lepiota' is more accurate. This, and other species with this epithet may appear reddish or even otherworldly, but they are not from Mars.

marz (L): March. *Hygrophorus marzuolus* is called the 'March Mushroom', and is hunted by bears and mushroomers alike.

mastoid (L): Breast-like. *Macrolepiota mastoidea* (mas TOY dee uh) is shaped somewhat like a breast. Related, 'mastodon' literally means 'breast tooth', the 'mastoid process' of the temporal bone of the skull looks like a little breast sticking out of the skull (maybe Galen was lonely when he coined the term), and a 'mastectomy' is the surgical removal of a breast.

mastrucat (L): Covered in wool. *Hohenbuehelia mastrucata* (mas true CAH tuh) has a similarity in appearance to sheepskin.

matsu (Japanese): Pine. The common name of *Tricholoma magnivelare* is 'matsutake' (mat sue TOCK ee), which literally means 'pine mushroom', due to its association with pine. A friend of mine used to say "Know Lodgepole, Know Matsutake. No Lodgepole, no Matsutake".

maurus (G): Dark, dark-skinned. *Mycena* and *Myxomphalia maura* (MORE uh) both have a dark colored pileus. Also, the Moors, a group of Muslims from North Africa, were named after this root as well.

medull (L): Marrow, pith. *Perenniporia medulla-panis* is the color of pith and bread. The medulla oblongata is a structure within the brainstem that utilizes this root.

mel (L): Honey. *Armillaria mellea* (mel LAY uh) is the 'honey mushroom', named because of the color of the mushroom.

melac (G): Soft. *Crepidotus malachius* (mal ACH ee us) is known as the 'soft-skinned *Crepidotus*'.

melan/mela (G): Black. *Geastrum melanocephalum* (mel AN oh ceph AL um) has a black head.

meleagr (G): A guinea-fowl, speckled. *Leucoagaricus* and *Cortinarius meleagris* (mel eh AG ris) have a resemblance to guineas. The European Guinea fowl is *Numida meleagris*, and the North American turkey is *Meleagris gallopavo*. Early Europeans incorrectly identified the North American turkey as Guinea fowl, thus leading to the European epithet being applied to the genus of the turkey.

melolontha (L): A genus of beetles. *Ophiocordyceps melolonthae* (mel oh LON thae) parasitizes *Melolontha* species. Many species, like this one, are named after their host.

membran (L): Skin of the body, parchment, thin. The epithets *membranaceum* (mem bran AE see um) and *membranacea* are used in the names of hundreds of fungi.

mephitic (L): Noxious-smelling, mephitic. *Russula, Tephrocybe, and Entoloma mephiticum* (mef IT ick um) all smell bad.

mer (G): Part. The genus *Meripilus* (mare ih PIL us) literally means 'part cap', presumably a reference to the fruitbody that has many parts of caps.

merd (L): Dung, excrement. *Deconica merdaria* (mer DAR ee uh) grows on dung. In the world of Spanish curse words, 'mierda' means 'crap'.

meridionalis (L): Southern. *Scleroderma meridionale* (mare ID ee oh NAL eh) is a southern *Scleroderma* species that typically grows in sandy soil.

merism (G): Part, division. The genus *Merismodes* was originally used because *Merisma* was taken. Both refer to the divided appearance of a fruitbody. The epithet of *Fusarium merismoides* means 'divided-looking'.

meruloid (mer OOH lee oid): Wrinkled with uneven ridges. *Byssomerulius* is a yellowish (Bysso-) crust fungus with uneven ridges.

mes (G): Middle. *Auricularia mesenterica* (mes en TER ic uh) is the wood-ear jelly fungus that looks like intestines (literally, 'middle intestine'). The center (middle) of the pileus of *Hebeloma mesophaeum* (me so PHAY um) is dark. This word element pops up in many applications, both in mushroom names and in common language.

metron (G): A measure. The specific epithet of *Astreaus hygrometrica* (hi grow MET rick uh) literally means 'water measurer'. They respond to water by opening up their rays to expose the gleba.

mica (L): To glitter or glisten. *Coprinellus micaceus* (my CAY see us) is the 'mica-cap Coprinus' because of the glistening specs on the pileus. As you may know, 'mica' is a glistening mineral, and 'Formica' was a replacement for mica.

Michener, Ezra (1794 - 1887) was a medical doctor with a passion for nature who was born and died in Pennsylvania. He frequently corresponded with Berkeley, Curtis, and Ravenel, and worked at the Schweinitz herbarium in Pennsylvania and contributed hundreds of specimens to it. *Lentinellus micheneri* was named after him.

micro (G): Small. Over 3,000 fungus epithets include the prefix 'micro' to denote something small about them.

militar (L): Belonging to a solider, warlike. *Cordyceps militaris* (mil ih TARE iss) attacks and kills bugs like the best of militaries.

milt (G): Red chalk. *Cortinarius miltinis* (mil TIN is) is reddish in color, and *Heterotextus miltinus* turns reddish with age.

milv (L): The genus name of the Kite. *Cortinarius milvinus* (MIL vin us) is the color of a kite's back.

miniat (L): Cinnabar, vermillion, red lead, a vivid red color. *Hygrocybe miniatus* (min ee AH tus) is the 'vermillion waxycap', and species like *Clavaria miniata* are reddish. The red color was used to denote paragraph signs, capital letters, and headings in manuscripts, and used in small drawings. Those doing this work were 'miniators', which led to the word 'miniature' as we know it. Indeed, most *miniatus* species are also small.

mirab (L): Marvelous, strange. Used in nearly 300 specific epithets, including *Aureoboletus mirabilis* (mir AB il is) and *Hydnellum mirabile.* Admirable is a related word.

miranda (L): Wonderful. *Strobilomyces* and *Cortinarius mirandus* (mere AN dus) are both beautiful. 'Admirable' is derived from the same word.

mitr (G): A head-band or cap. *Mitrula* (my TRUE luh) species have a little hat-like cap on the top of a stalk. In related words, the 'mitral valve' of the heart resembles a 'mitre', the head dress of Christian bishops.

mni (L): Moss. *Galerina mniophila* (mn ee AH fill uh) grows on moss. *Mnium* is a genus of moss.

mollis (L): smooth, soft. *Lycoperdon molle* (MOE ley) is a smooth puffball, and *Coltricia permollis* (per MOL lis) is soft all-over. In related words, animals in the phylum Mollusca are soft.

mollusc (L): Soft. *Hydnum molluscum* (mol LUS cum) resembles a mollusk. The epithet *mollis* also refers to something that is soft.

molybdos (G): Lead, lead-colored (a dull gray to a bluish-white). *Exidiopsis molybdea* (moe LIB dee uh) is the color of lead. Similarly, *Chlorophyllum molybdites* is supposed to be lead-colored. Apparently someone who has more experience with lead was naming these mushrooms.

Montagne, Camille (1784-1866): Montagne was a French physician and mycologist. He sent an interesting species to Fries, who described the specimen in 1836 and established the genus *Montagnea* in his honor. *Camillea* was also named after him. Montagne's biggest claim to fame was that he was one of the original scientists who described a fungus responsible for the potato blight, what we now call *Phytophthora infestans.*

morb (L): Disease. *Apiosporina morbosa* (mor BOE suh)'s common name is 'black knot disease'. I would prefer Kuo's description of 'cat poo on a stick' as the official common name, but 'black knot disease' is descriptive enough. 'Morbid' and 'moribund' are related words.

Morgan, Andrew Price (1836 – 1907): Morgan studied fungi in Ohio and studied under Curtis Gates Lloyd. He had several publications on the Gasteromycetes, and his name pops up several times in association with the puffballs/earthstars. *Astraeus morganii* is one mushroom named after him. The puffball genus *Morganella* is also named after him.

Morris, George Edward (1853 – 1916): Morris was the vice president of the Boston Mycological Club (the oldest mycology club in the United States!) and devoted much of his later life to studying mushrooms. He made over a thousand watercolor sketches of mushrooms, and had extensive correspondence with Charles Peck, who named *Boletus morrisii* after him in 1909.

morus (L): The genus of the mulberry tree. *Tremella (Pseudotremella) moriformis* is mulberry-shaped.

mucid (L): Slimy, musty, or moldy. Used in over 100 species names with the epithets *mucidum* (mu cid um), *mucidus*, and *mucida*. For example, *Crepidotus mucidifolius* has somewhat gelatinous gills. This word is related to 'mucus'.

mucr (L): A sharp point. The young caps of *Hygrocybe mucronella* (mu cron EL luh) are acutely conical, and members of the genus *Mucronella* have finely pointed teeth.

multus (L): Many. *Phallus multicolor* (MUL tee color) is a multi-colored mushroom, and *Polyozellus multiplex* (MUL tie plex) has many small branches. In related words, a 'multiplex' has many parts.

muricatus (L): Pointed, full of sharp points. *Elaphomyces muricatus* (MUR ih CAT us) is covered with pyramidal (pointed) warts.

murin (L): Mouse-like. *Amanita murinoflammeum* (mure in oh FLAM me um) has a mouse-colored partial veil. In related words, Murinae is the subfamily of rats and mice, 'murine' means 'related to mice', and *Mus musculus* is the binomial of the common house mouse.

Murray, Dennis (?) was a mushroom collector from Massachusetts. I can find nothing about him other than that he made noteworthy collections of fungi. Murray and Sprague (of *Fomitopsis spraguei* fame) collected 682 species in the 1850's, and sent them to Curtis. Many of the species were undescribed, and Curtis named them. The most well-known species named after Murray was *Agaricus murrayi*, which is now *Entoloma murrayi*. Sprague sometimes spelled 'Dennis' with a single 'n'.

Murrill, William Alphonso (1869-1957) was an American mycologist. He earned his PhD in 1897, was a curator for the New York Botanical Garden, and founded the journal *Mycologia*. He had over 500 publications, erected several new genera (you can thank him for *Suillellus*), collected tens of thousands of specimens, and formally described nearly 1,500 of them. Thirty species have been named after him, including *Tricholoma murillianum* (mur rill ee AN um), the matsutake from western North America (formerly called *Tricholoma magnivelare*).

murus (L): Wall. *Omphalina muralis* (mure AL is) grows on mossy walls. A related word is 'mural'.

musca (L): A fly. *Amanita muscaria* (muss CAR ee uh) was named because its traditional use as a fly killer (*Amanita muscaria* powder + milk + fly = dead fly). The main psychoactive compound in the mushroom is muscimol. Muscarine, a very different compound with very different effects, was originally extracted from this species and was named after the specific epithet. However, there is only a trace amount of muscarine in *A. muscaria*, and it does not contribute to the psychoactive effects of the mushroom.

muscus (L): Moss. The epithets *muscorum* (mus COR um) and *muscicola* refer to a mossy habitat and are associated with numerous genera. *Phaeocollybia muscicolor* is a mossy olive color. It is unrelated to 'muscaria', which means 'fly'.

mustach (French): Mustache. The cells on the gill edge of *Gymnopus mustachius* (mus TASH ee us) look like a brown mustache.

mustela (L): Weasel. *Russula mustelinicolor* (mus tel IN ih color) is the color of a weasel.

mutab (L): Changeable. *Russula mutabilis* (mu TAB il is) readily bruises. There are 221 other hits on Indexfungorum for this epithet describing species that change in some way, shape, or form.

mutinus (L): A name for Priapus, a Greek god of fertility who had a constant (and rather large) erection. It is used in *Mutinus* (mu TINE us), a genus of phallic-shaped mushrooms. In related terminology, 'priapism' is the medical condition of having an erection for a prolonged period of time.

myrio (G): Many, myriad. *Collybia myriadophylla* (mear ee ad oh FILL uh) (currently *Baeospora myriadophylla*) has many gills.

myrmec (G): Ant. *Ophiocordyceps myrmecophila* (mir meh CAH phil ah) love ants. Mymarommatidae is a family of hymenoptera, including the ants that the *Ophiocordyceps* love so much.

myrtil (L): Myrtle. *Cortinarius myrtillinus* is the color of myrtle, while species with the epithet *myrtillicola* grow in association with the plant.

myx (G): Slime. It is used in over 40 specific epithets to denote sliminess. It is also used in *Myxomycetes* (= *Myxogastria*), a taxonomic class of slime molds in the kingdom Protista. Perhaps the best known Myxomycete (mix oh MY seat) to mushroom hunters is *Lycogala epidendrum*, the wolf's milk slime mold.

N

naem / naima (G): Gelatine. The genera *Naematelia* and *Entonaema* are gelatinous. It is not used in the genus *Naematoloma*, which was a misspelling of *Nematoloma* (see –nem) and appropriately means 'threaded fringe'. *Psilocybe naematoliformis* is named after a likeness to *Naematoloma*, so that should probably be *P nematoliformis*, but at least it is partially gelatinous.

nameko (Japanese): Slimy. *Pholiota nameko* (nah MAY co) has a thin slimy covering on it, and it is also one of the most common cultivated mushrooms in Japan.

Nannfeldt, John Axel (1904 – 1985) was a well published Swedish mycologist who specialized in the ascomycetes, rusts, and smuts. *Helvella nannfeldtii* (nann FELD tee eye) was named after him. Additionally, the genus *Jafnea* takes the first letter of each initial, including the second middle name he obtained upon confirmation, 'Frithiof'.

nanos / nanus (G): Dwarfish, small. *Marasmius nanosporus* (nan oh SPOR us) has small spores, and *Tylopilus nanus* is small. In related words, 'nanotechnology' manipulates atoms (small indeed!), and a 'nanny' was traditionally a nurse for (small) children.

napus (L): Turnip. The base of the stem in *Cortinarius napus* (NAP us) is turnip-shaped.

narcot (G): Numbness, stupor. *Coprinopsis narcotica* (nar CAH tic uh) smells like opium. Rogers mushrooms didn't seem to associate the smell with opium, as their website describes the odor as "non-mushroomy" and "odd". I guess I'm with them in not knowing what the drug smells like.

nard (L): Nardus, spikenard, an oil from the plant Nardostachys. *Armillaria nardosmia* (nard OS mee uh) and *Puccinia nardosmiae* smell of the musky spikenard.

nauc (L): The shell of a nut. *Naucoria* (nau COR ee uh) species are reminiscent of nut shells.

nauseosus (L): Produces nausea. *Russula nauseosa* (naw zay OH suh) has a hot taste, and I am assuming someone had a bad experience with it.

nebulosus (L): Dark, clouded (color). *Geopyxis nebulosoides* (neb you low SOY dees) looks dark/cloudy. Related, a 'nebula' is a cloud of dust.

nectr (G): A swimmer. Berkeley stated in 1860 that the genus name *Nectria* is "in allusion to the fluxile contents of the perithecia", although I have no firsthand experience with this myself. Other sources (probably erroneously) state that the genus name is related to the Greek *necros*, allegedly a reference to the dead host of the fungus. *Necturus* is also a genus of salamanders, and anacondas are in the genus *Eunectes* (true swimmers). It is unrelated to the word 'nectar'.

neglect (L): Forgotten, overlooked, literally 'not chosen'. *Psathyra neglecta* (neh GLEC tuh) is easily overlooked. I had a picture of it, but I forgot where it went.

nem (G): Thread. The deprecated genus *Naematoloma* is named after the thread-like partial-veil or the thread-like rhizomorphs, depending on who you listen to. Interestingly, the new name for many of the old Naematolomas is *Hypholoma*, which also means 'thread-like' (think Hyphae). A related word is 'nematode', which are 'roundworms' or 'threadworms'.

nemus (L): Pasture or sylvan. *Coprinus nemoralis* (nem or AL is) and *Psilocybe nemophila* both grow in this habitat. *Nemoralis* is also the epithet of a common Goldenrod species.

neo (G): New. *Russula neoemetica* (knee oh ee MET ic uh) was a 'new' species (new in 1979, anyway), similar to *R. emetic*. Related words include 'neocortex' (new cortex) and 'neonatal' (new born).

nidus (L): Nest. *Nidularia* (nid oo LAR ee uh) is the 'bird's nest fungus'. A little obscure, but the word 'nidulate' means 'to make a nest'.

nigr (L): Dark, black. *Russula nigricans* (NIGH grih cans) is one of a number of black Russula species.

nitens (L): Shining, bright. *Gymnopilus purpureonitens* (pure pure eh oh NIT ens) has a bright shiny purple pileus.

nitid (L): Bright, glittering. *Scleroderma nitidum* (nih TIE dum) is bright.

niveus (L): Snowy. *Russula nivea* (NIV ee uh) and *Coprinellus niveus* are white as snow.

nobilis (L): Famous, known. *Russula nobilis* (no BIL is) and 77 other fungal species that use this element are generally stately mushrooms.

nola (L): A little bell. Members of the genus *Nolania* (no LAY nee uh or no LAN ee uh) are distinctly bell-shaped. Many (most?) *Nolania* species are now *Entoloma* species.

nota (L): Marked. *Cortinarius notatus* and *subnotatus* (sub no TAT us) are marked with fibrils. Nearly 50 other mushroom species contain a related epithet. Related words include 'note', 'notebook', 'banknote', 'footnote', and 'notary'.

noveboracensis (L): Literally means 'from New York' (novum 'new' + eboracum 'York' + ensis 'from'). *Rhodocybe noveboracensis* is the 'New York Rhodocybe'. There is also *Vernonia noveboracensis* (the New York Ironweed) and *Thelypteris noveboracensis* (the New York Fern). It is infrequently used as an adjective, but now you'll know what it is referring to the next time you encounter it.

novus (L): New. *Amanita novinupta* (no vih NUP tuh), the blusher from the western United States and Mexico is called the 'new bride blusher', because it appears to blush just under the surface of the mushroom.

nucleus (L): A little nut or kernel. *Exidia nucleata* (new clay AH tah) has the appearance of many kernels. In related terminology, 'a nucleus' is a collection of cell bodies in the brain, a little nut of tissue that shares a similar function.

nuda (L): Naked. Species like *Lepista* and *Ditiola nuda* (NEW duh) lack a veil.

nummul (L): A coin. *Collybia nummularia* (num mu LAR ee uh) is an example of a coin-shaped fungus. The moneywort, *Lysimachia nummularia*, is also coin-shaped.

nutan (L): Nodding. The head of *Ophiocordyceps nutans* (NUE tans) often droops down, as if by nodding. This species grows from stinkbugs.

nyct (G): Night. Used in many epithets in both plants and fungi to denote growth or appearance at night. Further, the word 'nychthemeron' refers to a 24-hour period of night and day. Thus, the epithet of *Coprinus nycthemerus* (nick THEM er us) refers to the mushroom's 24-hour life.

O

obes (L): Stout, fat, obese. There are 55 species that include *obesum* (oh BEE sum) or *obesa*, and they are all stout in one way or another.

obliquus (L): Slanting sideways. *Inonotus obliquus* (oh BLEE kiss) gets its epithet because of the slanting pores on the fungus. Knowing this fact alone would help many people who post pictures of a mushroom they think is chaga realize it is not chaga. In related terminology, you have some 'oblique' muscles.

obovat (L): With the narrower end at the base. *Microporellus obovatus* (ob oh VAH tus) has an obovate-shaped fruiting body.

obruss (L): Gold colored. *Hygrocybe obrussea* (ob RUS sea uh) is the color of pure gold

obscur (L): Dusky, dark. *Agaricus perobscurus* (per ob SCUR us) adds the prefix 'per', meaning 'all over' to the adjective describing the dark brown fruiting body. Related, 'obscure' means unclear or hidden.

obtectus (L): Covered over. *Russula obtecta* (ob TEC tuh) and 84 fungal species that use this epithet are all covered with something.

occidental (L): Western. *Russula occidentalis* (oc cih den TAL iss) is one of the more Western *Russula* species.

occult (L): Hidden. *Clitocybe occulta* (uh CUL tuh) and many other species got their names because they were hard to find or characterize. The 'occult' involves the supernatural, which is also hidden.

ocellus (L): Diminutive of oculus, eye. *Gymnopus ocellus* has an iris-like pattern on the pileus.

ochro (G): Ochre, yellow, pale. *Inonotus ochroporus* (oh kroe PORE us) is named because the color of the pore surface.

ocrea (L): A greave (a piece of armor between the knee and ankle) or legging. *Amanita ocreata* has a baggy shin guard-like volva.

odon (G): Tooth. *Hydnellum cyanodon* (cy AN oh don) has blue spines on its ventral surface. Related words include 'orthodontist' and 'endodontist'.

odor (L): Fragrant. *Clitocybe odora* (oh DOR uh) and others with this epithet have a pronounced odor.

odoratus (L): Sweet-smelling. *Hygrophorus odoratus* (oh dor AH tus) has a sweet odor.

oides (L): A likeness of form. *Lactarius hygrophoroides* (hi groph or OID ees) is like a Hygrophorus. This suffix is very similar to the suffix *opsis*, and just as common. Can you think of any other names that end in oides?

olea (L): The genus name of the Olive tree. Over 200 fungal species are named with reference to the olive tree, like *Omphalotus olearius*, the European 'jack-o'-lantern', which grows at the base of the tree.

olens (L): Smelling. The epithet *suaveolens* (sue ah veh OH lens) is used as a suffix in many odorous mushrooms. For example, *Lactarius cocosiolens* smells like coconut.

olid (L): Odorous, stinking. *Armillaria* and *Hygrophoropsis olida* (oh LIE duh) and two odorous mushrooms. *H. olida* smells like root beer.

oliva (L): Olive (color). *Cordyceps olivascens* (ol iv AY sens) literally means 'becoming olive'. It is a rare Cordyceps from Alabama and Mississippi, and it has a very light green to olive-buff stroma. It is in fact so rare that I cannot find a good picture of it on the internet.

oll / ollul (L): A pot / a little pot. *Cyathus olla* (AHL luh) and *Aleuria ollula* both resemble pots.

ombros (G): Rain storm. *Gymnopilus ombrophilus* (om BRAH fill us) loves the rain (but seriously, don't most?).

omphal (G): Navel, umbilicum (umbilicate). Mature specimens of *Omphalotus illudens* (om pha LO tus) have a depression in the center of the cap that looks like a navel. You can also use *umbilicate* as an adjective to describe a species that has a central depression in the cap. The next time you find a mature *Omphalotus*, squint your eyes, and you might just see a navel in the middle.

oniscus (L): The genus name of the woodlouse (pill bug, roly-poly). *Arrhenia onisca* (on IS cuh) resembles the color of the woodlouse. You have to wonder how soon this species was named after seeing a woodlouse.

ono (G): A donkey. *Otidia onotica* (on OT ic uh) literally means 'a little ear that looks like a donkey ear'.

onust (L): Loaded. *Amanita onusta* (on OOS tuh) is the 'gunpowder Lepidella' because the warts on the cap look like piles of gunpowder.

onych (G): 'Onychos' originally meant nails or claws, but was then applied to pink or yellowish minerals with white streaks. As applied in biological epithets, it typically means yellowish, the color of onyx marble. *Lyophyllum onychinum* (= *Tricholoma onychinum* and sometimes called *Rugosomyces onychinus*) (on ih KYE num) has onyx marble-colored gills. 'Onchoclasis' is the breaking of finger or toenails and an 'onychectomy' is the surgical removal of them.

ophi (G): Snake. *Elaphocordyceps ophioglossoides* (AH fee oh gloss OYE dees) (Formerly *Cordyceps ophioglossoides*) literally means 'snake tongue' because the tip of the fungus is often split in two at the end.

opsis (G): Appearance, similar to. This is suffix is commonly added to the name of a genus when it is split into multiple genera. For example, many members of *Coprinus* were split into *Coprinopsis* (co pry NOP sis). Can you think of any other names that in opsis?

or (G): Mountain. *Hebeloma* and *Steccherinum oreophilum* (or ay OPH il um) are from the mountains. Related is the epithet of *Marasmius oreades*, named after a nymph of the mountains.

or/os (L): Mouth. *Tricholoma orirubens* (or ee RU bens) has red-edged gills. This tiny word element is fit into many species names as well as common words (think 'oral').

orb / orbis (L): Disk, orb / circle. *Orbilia* (or BIL ee uh) species are shaped like little disks and *Ganoderma orbiforme* (OR bih FOR meh) is basically shaped like a circle. Related is the orbit of the eye.

oreades (G): In Greek mythology, nymphs of the mountains and hills. *Marasmius oreades* (or ay AHD ees) is called 'the fairy ring mushroom', but it is not the only species to reference these nymphs: *Russula oreades* is another.

orellana: Named after the plant *Bixa orellana*, itself named after the Spanish conquistador, Francisco de Orellana. The seeds of the plant are used to make annatto (also called achiote), which is used as both a spice and a food coloring agent. *Cortinarius orellanus* (or el LAN us) is the color of annatto.

orichalc (L/G): Literally 'mountain copper', it refers to a yellow metal, or a yellowish copper or brass color. *Cortinarius orichalceus* (or ih CAL see us) is yellowish, and *Cortinarius orichalceolens* becomes that color. Related is the epithet *chalciporus*, which has copper-colored pores.

ornat (L): Adorned, decorated, beautiful. *Retiboletus ornatipes* (or NAT ih peas) has a prominently reticulate (decorated) stipe.

ostre (L/G): Oyster. *Pleurotus ostreatus* (os trey AT us) is better known as the 'oyster mushroom', and *Stereum ostrea*, the 'false turkey tail' is shaped like an oyster too, kind of. Actual oysters (the bivalves) are in the family Ostreidae.

osus / osa / osum (L): Denoting augmentation, being full of. *Gymnopus spongiosus* (spongy OH sus) indicates that it is 'full of sponge' (very spongy).

ot (G): Ear. *Inonotus* (eye NON oh tus) literally means 'fibrous ear' and *Otidia* (oh TID ee uh) look like ear lobes. In related terminology, an 'otoscope' is a tool that looks in the ear, and 'otolith' is part of the inner ear.

Oudemans, Cornelius/Corneille Anton Jan Abraham (1825-1906) was a Dutch physician, botanist, and mycologist. He published three books on fungi, and the genus *Oudemansiella* (ooh de man see EL uh) was named after him in 1881.

ovis (L): After the genus name of sheep, *Ovis*. The sheep-polypore, *Albatrellus ovinus* (oh VIE nus) is the color of sheep.

ovoid (L): Egg-shaped. Frequently used as an adjective to describe spores, conidia, or cystidia. *Psilocybe ovoideocystidiata* (oh VOID ee oh sis tid ee AH tuh) has egg-shaped cystidia. I have also heard the phrase 'Ovoid hunter' to describe someone who specifically hunts this psilocybin-containing hallucinogenic mushroom.

oxy (G): Sharp, acidic. *Oxyporus* was named after its acidic taste. 'Oxygen' was thought to be required in all acids, and 'oxymoron' is a combination of sharp and foolish, essentially 'pointedly foolish'.

ozos (G): Branch or twig. *Polyozellus* (POLY oh ZEL us) features a branching fruitbody.

P

pachy (G): Thick. *Calvatia pachyderma* has thick skin.

Paden, John Wilburn: Paden was expert on Pezizales and a student of Tylutki. Indeed, it was Paden and Tylutki who formally described and named *Sarcosoma mexicanum*. His name appears in *Urnula padeniana* (pay den ee AN uh) (formerly *Sarcosoma mexicana*).

palae (G): Ancient. A prefix used in genera names and specific epithets. *Palaeoagaracites antiquus* was found in Burmese amber from the Cretaceous period and was around 100 Million years ago. Related words are 'paleontology' and 'Paleolithic'.

palea (L): Chaff. *Cortinarius* and *Irpex paleaceus* (pal ee AH cee us) are scaly.

pallidus (L): Pale, pallid. *Laetiporus gilbertsonii var. pallidus* (PAL id us) is similar to *L. gilbertsonii* but has white pores (i.e. is pallid), and is found in the Southern United States. It is found on oak and eucalyptus just like *L. gilbertsonii.*

palm (L): Palm tree. *Gymnopilus palmicola* (pal MIC oh luh) grows on palm logs.

paludis (L): A marsh, swamp, bog. *Russula paludosa* (pal ooh DOE suh) and other mushrooms with this epithet (and there are a lot!) grow in a marshy area.

palustr (L): Marshy. *Agaricus palustris* (pah LUS tris) thrive in marshy areas. 'Palustral' (pertaining to marshes) and 'palustrine' (wetlands without flowing water) are related ecology terms.

pan (G): All. Members of the genus *Panaeolus* (pan ay OH lus) are all over. This word element is used in too many words to list here, like 'pansexual' (loving everyone), 'pancreas' (all flesh), and 'pandemic' (all people).

pann (L): Rags, ragged. *Amanita cinereopannosa* (cin er ee oh pan NO suh) has a gray (ciner-) ragged margin.

pantherina (L): Panther, with spots like a panther. *Amanita pantherina* (panther EYE nuh) has warts (spots) on the cap.

panus (G): Swelling or tumor. The genus *Panus* (PAN us) is simply named after a resemblance to a swelling on the host.

papaver (L): Thick milk. *Russula papavericolor* (pah pay VER ih color) must have thick droplets. A more famous species that uses this word element is *Papaver somniferum*, the opium poppy. Here, the species name indicates that sleep comes from the thick milky substance (indeed, it contains morphine and codeine).

papill (L): Bump, pimple, nipple. *Otidia papillata* (pap ill AH tah) and many other species that have little bumps. In related words, 'papillae' are the small bumps on your tongue that contain your taste buds.

papyr (G): Stem of a plant used for paper, or generally, 'like paper'. 'Papyraceous' can be used as an adjective to describe something that is like paper, and is used in many species names, including *Stereum papyraceum* (puh pie RAH see um) and *Thelephora papyracea.*

para (G): Distinct from, beside, near. *Russula paraemetica* (pear uh ee MET ih cuh) resembles *R. emetic*, and *R. parahelios* resembles *R. helios*. In related words, the 'paranormal' is distinct from normal.

pard (G): Leopard. *Chlorociboria pardalota* (par doh LAT uh) has a leopard-like spotted pattern, and *Tricholoma pardalotum* is called the 'spotted Tricholoma' and 'leopard knight'. The Leopard itself is *Panthera pardus.*

parv (L): Small, petty. *Ganoderma parvulum* (par VIEW lum) is a small *Ganoderma*, as far as Ganoderma's go, anyway. In other terminology, the 'parvocellular' layers of visual structures is comprised of small cells.

pascu (L): Pasture. *Clitocybe pascua* (pas CUE uh) and *Entoloma pascuum* were originally described from pastures.

patell (L): A small dish or plate. Used in *Clitocbye patelliformis* (pah tell ih FOR mis), *Tectella patellaris* (formerly *Panus patellaris*), and over 100 other species. The 'patella' (knee-cap) is also like a small dish, and 'patina', a film that appears on a bronze dish, is also related.

patriot (L): Of one's own country, Patriotic. *Boletus patrioticus* has a red zone, and white flesh that turns blue.

patul (L): Spread out. *Leptotrema patulum* (pah TUL um) is named after its recurved pileus.

pauca (L): Few. *Marasmius paucifolius* (paw sih FOE lee uh) indicates the species has 'few gills'. *Lactarius paucifluus* essentially means it has 'a little latex'. In a related term, 'paucity' means insufficient. This is not to be confused with the Greek *pausis*, the origin of the word 'pause'.

pauper (L): Poor. Over 100 species have an epithet like *pauper* or *paupercula.* The specific meaning underlying the epithet varies, though. Some allude to growth in a poor habitat, another has the smell of feces, and there are probably other meanings too. An alternate title of Mark Twain's 'The Prince and the Pauper' could have been "*Agaricus augustus* and *A. pauperatus*".

pausiac (L): Olive colored. *Amanita* and *Clitocybe pausiaca* are both olivaceous.

paxillus (L): A peg, small stake. Members of the genus *Paxillus* (pax ILL us) and species like *Russula paxilloides* have a resemblance to stakes in the ground.

Peck, Charles Horton (1833-1917): *Hydnellum peckii* (PECK ee eye) is named after him. He described thousands of new species and wrote several books. His detailed 'Annual Reports of the New York State Botanist' publications from the late 1800's and early 1900's are classics.

pectin (L): A comb. The margin of *Russula pectinata* is sulcate like the teeth of a comb. *Gaestrum pectinatum* is called the 'comb-like shell puff' in the older British literature, apparently a reference to the cuts in the exoperidium which form the rays. In broader use, 'pectinate' is an adjective used in zoology to describe a comb-like structure.

pediades (L): Of the plains. *Agrocybe pediades* (ped ee AD ees) grows in grass.

pedis (L): Foot. *Cantharellus altipes* (AL tih peas) is a species that was described in 2011 growing with oak and pine in Texas. As you might expect, it has a rather long stipe.

pelargonium (G): The genus name of the Geranium. *Russula pelagonia* (pel ah GO nee uh) smells like geranium flowers, while *Puccinia pelagonii* is a pathogen on the plants.

peli (G): Black and blue. The base of the stem of *Amanita pelioma* (pel ee OH muh) bruises a dark blue.

pellucidus (L): Transparent. *Auricularia semipellucida* (semi pel LOO cid uh) is partially transparent.

peltatus (L): Armed with a shield. *Auricularia peltata* (pel TAH tah) looks like little shields, and the common name is the 'shield fungus'.

pendul (L): Hanging down. The spines of *Irpicodon* (formerly *Irpex*) *pendulus* (PEN du lus) hang down. A related word is 'pendulum'.

penetrans (L): Penetrating. *Gymnopilus penetrans* (PEN eh trans) is also called the 'penetrating Gymnopilus'. It erupts from conifer logs and debris.

penion (G): Shuttle. *Peniophora* (pen nee AH for uh) has shuttle-shaped cystidia. *Peniophora* was first published in 1879, so we aren't talking about space shuttles. A shuttle is a wood piece that holds a bobbin, which is kind of shaped like a space shuttle. The second part of the species name, *-phero,* indicates that the hymenium bears the cystidia.

per (L): A prefix meaning all over, throughout. *Coltricia permollis* (per MOL lis) means 'soft all over'.

Perceval, Cecil H. Spencer (1849 - 1920): Perceval was a founding member of the British Mycological Society and made numerous collections in the late 1800's. He found and described *Battarea phalloides*, and *Leratiomyces percevalii* was named after him by Berkeley & Broome. There are no known pictures of him.

perdon (G): To break wind (flatulence). *Lycoperdon* (LIKE oh PER don), the genus name of several common puffballs, literally means 'a wolf's fart'. No one will ever say that mycologists aren't quirky.

perennis (L): Throughout the year, everlasting. *Coltricia perennis* (pear EN niss) lasts throughout the year. The word 'perennis' is also on United States $50 notes from the late 1700's: It appears over a 13-step pyramid.

pergamen (L): Parchment, which was originally from Pergamum. *Lactarius pergamenus* (per GAM en us) and *Trametes pergamena* are named after their resemblance to parchment.

perlat (L): Widespread. *Lycoperdon perlatum* (per LAT um) is a widespread fungus.

peronatus (L): Rough-booted, sheathed. The base of the stem of *Gymnopus peronatus* (per on AT us) is woolly, and the base of the stem of *Agaricus subperonatus* is buried. In general, 'peronate' can also be used as an adjective.

perplex (L): Perplexing, confused. *Hypholoma perplexum* (per PLEX um) was confused with other *Hypholoma* species.

persica (L): Peach. *Cantharellus persicinus* (per SIC in us) is peach-colored, and it was extra appropriate because this chanterelle smells fruity. There are 225 results from a wide range of genera for 'persic' in Indexfungorum.

personat (L): Mask. *Lepista personata* (per son AH tuh) (=*Tricholoma personatum*) is called the 'masked Tricholoma' because of the color changes the cap undergoes. 'Persona' is a related word.

pes (L): Stem. The stem of *Russula farinipes* (fair IN ih pees) is covered in a flour-like powder.

petaloid (E): Shaped like a flower petal or leaf. The individual fruit bodies of *Hohenbuehelia petaloides* are shaped like individual flower petals.

petasat (L): A hat used for travelling. *Pluteus petasatus* (pet uh SAT us) has a pileus that looks like a hat. As Tom Volk said, "You have to have a good imagination to be a mycologist". A Latin version of "The Cat in the Hat" is titled "Cattus Petasatus".

peziza (L): Sessile, lacking a stalk (as with puffballs, earth stars, and some other fungi). Members of *Peziza* (peh ZEE zuh) and the family Pezizaceae generally lack a stalk.

phae (G): Dark, dusky, gray (color). *Phaeolus* (fay OH lus), *Russula phaeocephala* (fay oh ceh PHAL uh), and the genus *Phaeocollybia* all contain references to a dark feature.

phaner (G): Visible. Used in *Panus hygrophanus* (hi GROH phan us). Have you ever noticed a mushroom cap where the outer part of it is a different color than the inner part? If so, you have probably noticed a hygrophanous cap. This is caused by differences in water absorption across the cap (notice the word starts with the prefix 'hygro', meaning 'water'). Darker rings of the cap are areas that retain water and are transparent. Conversely, lighter rings on the cap are areas that are drier and appear opaque.

phasma (G): An apparition, a ghost. *Cantharellus phasmatis* (fas MAT is) was described and named by Foltz, Perez, and Volk (2013). It is named after "the distinctive ghostly white hymenium of young specimens of this chanterelle". Also, the genus *Fantasmomyces* has ghost-like ascomata.

phellos (G): Cork. *Phellinus* (phel LIE nuss) are cork-like.

phila (G): Loving. *Russula hydrophila* (hi DRAH fill uh) is attracted to habitats with water like a fly to a stinkhorn.

phleb (G): Vein. Members of the genus *Phlebia* (FLEE be uh) look very vein-like. In related terms, 'phlebotomy' involves the cutting (-tomy) of veins to take blood samples.

phlog (G): A flame, reddish. There are over 50 epithets like *phlogis* (FLO gis) and *phlogina* that are a reference to a reddish color. The word 'phlogistic' means 'related to a fever'.

phoenic (G): Red, scarlet, reddish-purple. Many species in the fungal and plant world utilize the color-referencing epithet *phoeniceum* (phuh NEE cee um) or phoenicea. 'Phonecians' are from 'Phonecia' and that literally means 'purple country', which is named after the purple fabric they dyed from a local species of mollusk. The reddish-purple 'Phoenix' was also named after the color.

phol (G): Scale. *Pholiota* (foe lee OH tuh) typically has pronounced scales.

phora (G): Bearing. The name of the genus *Thelephora* (thel EPH or uh) literally means 'nipple-bearing'. Can you think of another fungus that utilizes the ending 'phora'?

phryg (G): Pertaining to Phrygia, where king Midas was from, thus, gold-colored, or fringed in gold. *Cortinarius phrygianus* (frige ee AN us) is gold colored.

phthor (G): Destruction. *Phytophthora* (fie TOPH thor uh) *infestans* is destructive to potatoes. 'Phthor' is also an old name of the element 'Florine', an explicit reference to its destructive properties. Florine, the name that eventually won, is from the Latin 'to flow'.

phyt (G): Plant, tree. *Phytophthora* (fie TOPH thor uh) *infestans*, is the name of the fungus that causes potato blight. Related words include 'phytobiology' (the biology of trees), 'saprophyte' (for example a mushroom living on dead organic material like a tree), and 'bryophyte', which literally means 'lichen plant'.

pica (L): The magpie. *Coprinopsis picacea* (pie CAY see uh) has white patches on the pileus that are reminiscent of those on the magpie (genus *Pica*). A related word, 'pica', refers to indiscriminate eating of non-food items. Although the etymology of this meaning is unknown, some believe it is from the indiscriminate eating habits of the magpie.

picea (L): Genus name of the spruce tree, also means pitch pine, from 'pix', itself meaning 'tar black'. *Cerospora piceae* grows on spruce and *Amanita picea* is known as the 'pitch black Amanita'. It is impossible to know whether species with these epithets grow with spruce or are named because they are black based on their name alone. Further, there are some species named after the magpie (genus *Pica*), like *Coprinopsis picacea*, which are not necessarily black or grow with spruce.

picin (L): Pitch-black. Members of *Picipes* (PICE ih pees) have a black stem. The best known example is *Picipes badius* (formerly *Polyporus badius*).

picr / pikros (G): Bitter. *Gymnopilus picreus* (PIC rhee us), like many other *Gymnopilus* species, is bitter.

pict (L): Pained. *Suillus pictus* (PIC tus) is called the 'painted Suillus'.

pil / pilus / pilos (L and G): Hair. *Russula pilocystidiata* (pie loh cis tid ee AH tuh) and *R. pilosella* have features (like cystidia) that are reminiscent of hairs. In related words, you might be familiar with 'piloerection', which is what happens when a cat or dog is scared and their hair stands up on end.

pinax (G): Picture, more generally, a likeness. *Dacryopinax* (dac rhee oh PIN ax) are shaped like tears.

pinguis (L): Grease, fat. The cap of *Cortinarius pingue* (PING gay) (formerly *Thaxterogaster pingue*) is greasy.

piper (L): Pepper, while 'piperatus' specifically means 'peppery'. *Lactarius piperatus* (pip er AH tus) and *Russula piperata* have a peppery taste. It might take a few seconds to develop and last for several minutes.

pipt (G): To fall. The pores easily fall off of *Piptoporus* (pip toe POR us) *betulina* (this is now called *Fomitopsis betulina*, but the name is still prevalent enough in the literature to include here).

pis (G/L): Pea. *Pisolithus* (pis oh LITH) has pea-shaped peridioles inside. *Lycoperdon pisiforme* (pee sih FOR may) is pea-shaped, while the epithet *pisicola* is used in a number of species that grow on the pea plant, which itself is represented by the genus *Pisum*.

pisc (L): Fish. *Lactifluus* and *Tylopilus pisciodorus* smell like fish. *Exophiala pisciphila* is a fungus that grows on fish. The zodiac symbol of 'Pisces' is fish, and a 'pescatarian' eats fish.

pistill (L): Pestle, the tool used for grinding things in a mortar. Used in long tapering pestle-shaped species like *Clavariadelphus pistillaris* (pis til LAR is). The 'pistil' of a flower is also named after its resemblance to a pestle, but 'pistol' (the gun) is unrelated.

pity / pithy (G): Pine. *Clitocybe pithyophila* was called the 'pine-loving Clitocybe' (until its name changed). Still, there are many other *pithyophila* and *pityophyla* species out there.

plac (G): A flat cake or a flat plate. *Agaricus placomyces* (plac oh MY cees) is a flat mushroom (at maturity, anyway). A related word, 'placenta', is also flat and round, and is named after a plate, not a cake (but to each their own).

placid (L): Pleasing, peaceful, placid. I can think of more pleasing fungi, but *Suillus placidus* (PLAH cid us) is up there.

plan (L): Flat. *Collybia planipes* (PLAN ih peas) has a flat foot (base of the stem), *Agaricus planipileus* has a flat cap, and *Graphyllium planispora* has flat spores. Related words include 'plane' and 'planar'. 'Plan' in the Greek sense means 'roaming or wandering' – think of the world 'planet'. It is unclear whether 'airplane' derives from the Latin or Greek since both make perfect sense. I believe all the fungal names that use 'plan' refer to physical flatness, including the epithet in *Ganoderma applanatum.*

platy (G): Flat, broad. *Otidea platyspora* (plat ee SPOR uh) has flat/broad spores. In related terminology, the platypus has a flat tail.

plect (G): Plaited, twisted, or wreath (which is coiled). There are a lot of cross-veins in *Mycena plectophylla*, and the name of the genus *Plectania* (plec TAN ee uh) appears to refer to the twisted base of the stem that goes into the substrate. Finally, species with the epithet *plectranthicola* grow on *Plectranthus* species, which are twisted themselves.

pleur (G): Side. Used in combination with *ot* (meaning ear) in the genus *Pleurotus* (plur OH tus), which can look like an ear on the side of a tree. Think about THAT next time you cut them off a tree.

plexus (L): Comprising a number of parts. *Polyozellus multiplex* (MUL tie plex) means 'many small branches, many parts', referring to the compound fruit body. In related words, a duplex has two parts, a triplex has three, and the Googleplex is the headquarters of Google.

plic (L): To fold. *Parasola plicatilis* (plic uh TIL is) gets posted almost daily in the late spring and summer on Facebook. I love how the cap on this one truly looks pleated.

plinth (L): Brick. *Lactarius subplinthogalus* has brick-colored milk. Although the milk itself is white, it dries and stains the mushroom tissue pink.

ploran (L): To cry or tear. In Russula section Plorantes (plor AHN tays), young specimens can form droplets of water on the gills.

plum (L): Feather. *Inocybe plumosa* (plume OH suh) was named after the downy tufts they have on their caps. A related word is 'plumage'.

plumb (L): Lead, lead-colored (bluish-grey). *Tylopilus plumbeoviolaceus* (plum bay oh vie oh LAY see us) is a lead-violet color. In related words, 'plumbers' and 'plumbing' traditionally used lead, and the abbreviation of lead on the periodic table of the elements in Pb.

plumig (L): Feathers, plumage. The epithet in *Cortinarius plumiger* (PLOO mih ger) literally means 'bearing scales'.

pluteus (L): The etymology of *Pluteus* (PLOO tee us) is invariably described as "difficult to pin down". Depending on the source, the genus *Pluteus* gets its name because of a resemblance to a shed, a penthouse, or to an ancient Roman 3-wheeled wooden frame used for protection as one advances on their opponents in war. This is not to mention the Greek meaning: wealth or riches. The most plausible explanation I've seen is from 'British Basidiomycetes' by W.G. Smith in 1908. The name Pluteus is "from a fancied resemblance in the pileus to the roof of a turret or sentry box".

pluv (L): Rainy. *Cortinarius pluvius* (PLOO vee us) is viscid and has watery flesh (as well as a watery taste, whatever that means). 'Pluvial' is a geological term that pertains to rain, and a 'pluviophile' describes someone who takes thoroughly enjoys the rain.

pocill/pocul (L): Cup, bowl. *Helvella pocillum* (PAH cil lum) is shaped like a cup, while the base of *Agaricus pocillator* forms a shallow cup. The epithet *pocilliformis* (pah cil ih FOR mis) is used in several species and specifically means 'cup-shaped'.

pod (G): Foot. This is used in countless specific epithets and even the genus *Podaxis* (poe DAX is). Related words include 'podium', 'Gastropod', and 'Cephalopod', to name a few.

pogon (G): Bearded. Members of *Rhizopogon* (rye zoe POE gone) have beard-like rhizomorphs, and some of the beards get a little unruly. 'Pogonophobia' is the fear of beards and 'pogonophilia' is being really, really into them.

poli (G): Grey. Used in over 100 species, including *Entoloma poliopus* (POE lee oh pus), which has a gray stem. Polio, the disease, is short for 'poliomyelitis', named after the grey matter in the spinal cord that degrades with the disease.

polit (L): Smooth, polished. *Entoloma polita* (POE lit uh) and other species with this epithet are named after their smooth and polished-looking surfaces. A related word is 'polite', but 'politics' is clearly unrelated. The epithet *impolit* means 'not smooth', or 'rough'.

poly (G): Many. *Polyozellus* (POLY oh ZEL us) has many branches. Poly is a common prefix in biology and binomial names in particular. For fungi alone, over 2,500 species start with it.

pom (L): Fruit tree or fruit, especially an apple. *Cytospora pomicola* (pom ih COE luh) is a pathogen on apples, *Aryria pomiformis* is a fruit-shaped slime mold, and *Phellinus pomaceus* is a pathogen on *Prunus*. 'Pomona' was the Roman goddess of fruit, a 'pome' is a fruit, and the word 'pomegranate' literally means 'apple with seeds'.

ponderosa (L): Weighty, ponderous. *Tricholoma magnivelare*, known as 'matsutake' in Japanese, was called *Armillaria ponderosa* (pon der OH suh) in 1887 and then changed to *Tricholoma ponderosum* in 1951 (and I still meet people who call it by these older names!). The epithet p*onderosum* appears that it was likely used because it grows in association with *Pinus ponderosa* (Ponderosa Pine), which was itself named because of its heavy wood. 'Ponderous' means 'slow and clumsy because of great weight'.

populus (L): Popular tree. *Amanita populiphila* (pop you lih FILL ah) loves poplar trees. Although found under aspens and cottonwoods, I have only found them under aspen stands.

porphyr (G): Red-brown. The genus *Porphyrella* (por fir EL uh), and species with related epithets, are red-brown.

porrig (L): Horizontally stretched out. *Pleurocybella porrigens* (POR rih gens), 'angel wings', has a stretched out cap. It is unrelated to 'porridge'.

portentosus (L): Monstrous. *Laetiporus portentosus* (por ten TOE sus) is rather large (up to 350 mm).

portobello: The word 'portobello', a common name of a large commercial Agaricus species, was coined in the late 1980's or early 1990's. Despite how recent this term is, there is still no clear evidence of where it came from. The best guess is that it is from the Italian word 'pratarolo', which means 'meadow mushroom'.

portoricense (L): From Puerto Rico. *Portoricense* is to Puerto Rico as *cubensis* is to Cuba and *costaricense* is to Costa Rica. I'll take Latinized forms of Latin

American countries for a thousand, Alex. There are over 125 hits on Indexfungorum for *portoricense*, including *Dentocorticium portoricense.*

postal (L): Place. Used in the genus *Postia* (POS tee uh). Related is the 'post' and 'post office'.

Pouzar, Zdenek (1932 - current): Pouzar was a Czech mycologist, and long-time editor-in-chief of the jourrnal *Czech Mycology*. The subgenus/genus *Pouzarella* (poe zar EL uh), related to the genus *Entoloma*, was named after him in 1976.

prae (L): Before, early, in front. *Gymnopilus praecox* (PRAE cox) comes early in the year. *Agrocybe praecox* is the 'spring Agrocybe'. 'Precocious' is derived from the same word.

prason (L): A leek. There are over 150 hits in Indexfungorum that include epithets using this word element. For example, *Cortinarius prasinus* (PRAS in us or pra SIN us) is named for the color-similarity to the leek.

praten (L): A meadow. The epithets *pratensis* (pray TEN sis) and *pratense* are used in over a hundred species names. *Pratense* is also the epithet used for some grasses.

prav (L): Irregular, deformed. *Morchella pravus* (PRAV us) has an irregularly shaped fruiting body. 'Deprave', meaning 'thoroughly crooked' is related.

primula (L): Primrose, which is a light pink color. *Marasmiellus* primulinus (prim you LIE nus) is primrose-colored, and *Tricholoma primulibrunneum* is primrose-brown.

prin (G): The evergreen oak. *Russula prinophila* (prin AH fill uh) grows in association with this tree.

privig (L): Stepson or stepdaughter. *Amanita privigna* (pri VIG nuh) is called the 'Step daughter death cap' because of its similarity to the better-known death cap. *Cortinarius privignoides* also has a likeness to a similar species.

procer (L): Tall, stretched out. *Lepiota procera* (pro CER uh) is tall and has a fairly long stipe.

prolix (L): Long, extended. Many species, including *Pholiota*, *Rhodocollybia*, and *Imbricaria prolixa* (pro LIX uh) are long in some way. Species with the epithet *prolixula* are only a little long. If someone is long-winded, they are displaying 'prolixity'.

propinquus (L): Near, neighboring, proximity. *Otidia propinquata* (pro ping QUAH tuh) grow close to one another in small clusters. A related (and probably obscure) word is 'propinquity', which means 'proximity'.

prot (G): First, primary. A prefix used in many genus names as well as specific epithets. For example, *Protomycena* was found in amber from the Eocene era, and was around approximately 40-million years ago. Related words include 'Protista', 'prototype', and 'protein'.

protract (L): To draw out. *Collybia protracta* (pro TRAC tuh) has broad (protracted) gills that are 6 mm thick! 'Protractor' is related.

pruinosus (L): Powdery, frosted. *Polydesmia pruinosa* (pru in OSE uh) and numerous other species that have a similar ending, have a powdery appearance. 'Pruinose' is a useful adjective to describe a powdery appearance.

prun (L): A plum. *Clitopilus prunulus* (PRUN you lus) resembles a little plum tree with a white bloom covering it. Other species include *Boletus prunicolor*, *pruniformis*, and *pruniodora* which refer to a mushroom's color, shape, and odor.

psamm (G): Sand. *Russula psammophila* (psam AH fill uh) grows in sand. This word element is used extensively, and gets nearly 100 hits on Indexfungorum.org.

psathyr (G): Brittle, friable. This is a synonym of the Latin *frio*. *Psathyrella* (sath er EL uh) are often called Snap-er-ellas (by my friends, anyway), because the stem often makes a snapping sound when it is broken in half.

pseudes (G): False, deceptive. P*seudohydnum* (sue doe HID num) is a monotypic species that deceptively looks like a Hydnum. A pseudonym is a false name a writer assumes.

psil (G): Naked, smooth. The genus name *Psilocybe* literally means 'smooth head'. An active chemical in Psilocybe mushrooms, psilocybin, was coined by Albert Hofmann. Species with the epithet *psilodermus* have a smooth surface, *psilophylla* have smooth gills, *psilpus* have a smooth stem, and *psilospora* have smooth spores.

psitt (G): Parrot (reference to color). *Gliophorus psittacinus* (formerly *Hygrocybe psittacinus*) (sit uh SIN us) is green like a parrot and almost as slimy as a hagfish when they are young.

pter (G): Wing. The genus *Pteridiosperma* (tare id ee oh SPER muh) has wing-like ridges on its ascospores. This word element is also used in 'pterosaur', the group of winged-dinosaurs.

pterid (G): Fern. *Mycena pterigena* (pter ih GEN uh) grows on dead fern stems, and the many species with the epithet *pteridicola* grow on ferns too. 'Pteridology' is the study of ferns.

pubescen (L): Hairs of puberty, downy. *Russula pubescens* (pue BESS cens), *Lactarius pubescens*, and nearly 200 other fungal species have fine hairs.

Puccini, Tommaso/Tomaso: Pier Antonio Micheli (1679-1737) named a group of rust fungi *Puccinia* (pue CIN ee uh) after Tommaso (or Tomaso) Puccini in 1729. Reports indicate that he was a professor of philosophy and/or anatomy in Florence, although I can find no details on why it was named after him. Several sources on the internet point to an Italian Wikipedia page about Tommaso Puccini (1749-1811) when discussing the genus *Puccinia*. However, this Puccinni was born 20 years after *Puccinia* was named, so it couldn't have been him. One Tommaso Puccini lived from 1666-1726, but another source definitively states the name-sake of the genus died in 1735. Unfortunately, the limit of what we seem to know about *Puccinia* is that it was named after an Italian in Florence who died a long time ago.

pudorinus (L): Pink, blushing. The pileus of *Hygrophorus pudorinus* (pew dor EYE nus) is pinkish-tan.

puell (L): Girlish: small, tender, and gracile. *Russula puellaris* (poo el LAR iss) got its name because it is slender and fragile. They get a pass on promoting gender stereotypes because it was named in 1838. *Cortinarius puellaris*, on the other hand, was named in 2015.

pulchell (L): Beautiful. *Russula pulchella* (pul CHEL luh) and *R. pulchralis* are beautiful fungi indeed.

pulchr (L): Beautiful. *Gymnopilus pulchrifolius* (pul chri FOLE ee us) has beautiful gills.

pullus (L): Dark, dusky, a reference to color. *Mycena pullata* (pul LAH tuh) is a dark-umber.

pululahuana (Quechuan): From Pululahua, Ecuador. It literally means 'cloud of water', because of the mist that rises on the volcano of the same name. *Ductifera pululahuana* was named after the location in Ecuador where it was first discovered. Pululahua is pronounced (poo loo LAH wah) in Quechuan, so the best we can do for pululahuana is 'poo loo lah WAH nah'.

pulverulentus (L): Dusty. *Russula luteopulverulenta* (loo tay oh pul ver you LEN tuh) looks like it is colored with a yellow dust. In related terminology, the word 'pulverize' means to reduce to powder or dust.

pulvinatus (L): Cushion-shaped. *Ophiocordyceps pulvinata* (pul vin AH tus) has stromata that are shaped like a cushion.

pumil (L): small. *Russula pumila* (pue MIL uh) is a small Russula.

punctatus (L): Spotted as with punctures. *Fomitiporia punctata* (punc TAH tah) is spotted. There are over 200 hits on Indexfungorum for *puncata* and *punctatus*, surely a lot of spotted mushrooms out there!

pungen (L): Pungent. *Russula, Entoloma, Suillus, and Lactarius pungens* (PUN gens) are all malodorous.

puniceus (L): Reddish, purple. *Exidiopsis punicea* (pew nih SEE uh) is a reddish color. This word element is common in many other genera to describe a fruit-body's color.

punk (E): After the punk hairstyle. The cap of *Mycena punkissima* looks like a punk.

purpurea (L): Black. *Russula atropurpurea* (AH tro pure PURE ee uh) has a black (*atro*) to purple (*purpurea*) cap.

pus (G): Foot, or stem as it pertains to a mushroom. *Mycena haematopus* (he ma TOE pus) exudes a blood-red juice from a crushed stem. *Russula erythropus* has a distinctly red stem, and many other mushrooms utilize this word element too.

pusillus (L): Very small or petty. *Gymnopilus pusillus* (puh SIL lis) and *Exidia pusillus* are both small. There are 101 records on Indexfungorum for this epithet.

pustul (L): Blister, pimple. Used in the genus *Pustula* (pus TOO luh) and many species like *Boletellus* and *Cortinarius pustulatus*. In English, a 'pustule' is a pus-containing pimple.

putid (L): Rotten, fetid. *Phellodon putidus* (PEW tid us) and *Russula putida* are malodorous.

pycn (G): Dense. *Pycnoporus* (pick no POR us) has dense pores.

pygm (G): Dwarfish. *Phallus pygmaeus* (pig MAE us) gets to 1 cm high when fully expanded. This is yet another micro-phallus.

pyr (G): Fire. *Lactarius pyrogalus* (pie rho GAL us) means 'fire milk'. 'Pyre', 'pyrotechnics', and 'pyromaniac' are related words.

pyr (L): Pear. *Lycoperdon pyriforme* (pier ih FOR meh) is shaped like a pear, and *Inocybe pyriodora* smells like a pear.

pyren (G): Pit, kernel. The Pyrenomycetes look like little pits, and the nearly 250 specific epithets use this root too, like *pyrenophora*, meaning 'bearing pits'.

pyrrh (G): Flame-colored, reddish. *Nectria pyrrhochlora* (peer uh CHLOR uh) and *Coprinellus pyrrhanthes* have a reddish color. The phrase 'pyrrhic victory', a victory where the losses are as great as those of the opponent, is somewhat-related. That phrase comes from the ruler Pyrrhus of Epirus, who experienced a costly victory and said 'If we are victorious in one more battle with the Romans, we shall be utterly ruined'. A related word is 'pyrotechnic'.

pyxid (L): A small box. The small branch-like tips of *Atromyces pyxidatus* (formerly called *Clavicoronia pyxidata*) (pix ee DAT us) look like little boxes.

Q

quad (L): Four. Mature fruiting bodies of *Coprinus quadrifidus* can split into four sections. *Cortinarius quadricolor* has four colors.

Quelet, Dr. Lucien (1832 - 1899) was a French naturalist, author, and mycologist. He was the first president of the French Mycology Society (Société Mycologique de France) and his name is used in *Boletus* and *Russula queletii*. He described several very common species, including *Agaricus bitorquis* and *Clavariadelphus truncatus*. I have commonly heard the epithet pronounced 'cue LET ee eye', but I believe this is a somewhat Americanized version.

quercus (L): Oak. *Gymnopus quercophilus* (quer coe PHIL us) grows on Oak leaves. Since the second half of the species name is *philus*, you can say that the mushroom loves oak leaves.

quietus (L): Mild, quiet, calm. *Lactarius quietus* (KWHY eh tus) references the drab coloring of the pileus. *Inocybe quietiodor* smells like *I. quietus.*

quintocentenaria (L): Quincentenary. *Torrendiella quintocentenaria* (quin toe cen ten AIR ee uh) was named to celebrate the 500th anniversary of Cristobal Colon (Christopher Columbus) landing in The Americas. On October 12th, 1492, Colon landed in what is now the Bahamas, and 500 years later, this ascomycete was named to commemorate it.

R

racemus (L): A cluster, or the stalk of a cluster. Epithets *racemosa* (rah cee MOE suh) and *racemosum* refer to general growth patterns, or, in the case of *Clitocybe racemosa*, for the fine hairs clustered around the stem. A related word is 'racemic', a word used in chemistry to denote a mixture of an equal amount of left and right handed molecules, which was first applied to a cluster of grapes.

radic (L): Root. *Xerula radicata* (rad ih CAH tah) *radicata* has a tap root.

radul (L): A scraper, small teeth. *Polyporus radula* (RAD you luh) has comb-like pores. *Radulomyces* is a genus of toothed crust fungi, and *radula, radulans,* and *raduloides* are used in over 50 epithets. 'Radula' are the teeth-like structures of mollusks.

rament (L): A shaving, scale, or chip. *Tricholoma ramentaceum* (rah men TAY see um), and many other species with this epithet have scales on the pileus.

ramus (L): Branch. *Ramaria* (ruh MARE ee uh) is a genus or nicely branching coral mushrooms, and *Laterispora brevirama* has short branches. A related word includes 'ramified', which means having branches or offshoots.

rancid (L): Stinking, rancid. *Clitocybe* and *Tephrocybe rancida* are malodorous.

rang (Chinese): Hymenium. The genus *Baorangia* (bow RANG ee uh) includes some boletes that have a thin hymenium.

rav (L): Gray. *Galera ravida* (RAV id uh) is gray, as are species with the epithets *ravida, ravidus,* and *ravidulus.*

Ravenel, Henry William (1814-1887): Ravenel of South Carolina was the first to describe *Phallus ravenelii* (rav en EL ee eye). This one is purported to routinely get up to 20 cm (8 inches) high, but we all know that mushroomers and hunters exaggerate lengths.

rech (G): Rough. The spores of *Scutellinia trechispora* (trech ih SPOR uh) appear rough.

recisus (L): Cut back, or having a short base. *Exidia recisa* (reh SEE suh) has a pseudostem that is somewhat tapered in appearance.

redolens (L): Redolent, fragrant. *Russula redolens* (red OH lens) has a pronounced odor. Redolent means strongly reminiscent of something, but also means fragrant or sweet-smelling.

regal (L): Royal. *Amanita regalis* (rhe GAL is) is a stately mushroom from Europe (it is also found in Alaska), and is nick-named the 'royal fly Agaric, and 'king of Sweden'. 'Regal' means 'fit for a king'.

regin (L): Queen. *Boletus regineus* is called 'the Queen Bolete'. Regina is the capital of Saskatchewan, which is named after Queen Victoria.

renes (L): Kidneys. *Marasmius*, *Crepidotus*, and *Hymenochaete reniformis* (ren ih FOR mis) are all kidney-shaped. 'Renal' is commonly used to refer to the kidneys.

repandus (L): Bend up, recurved. *Phellinus repandus* (reh PAN dus) and *Hydnum (Dentinum) repandum* are both recurved.

repraesent (L): Represented, well-represented. *Lactarius repraesentaneus* (reh pray sen TAY nee us) is widely distributed and can be found growing in groups.

resina (L): Resin. *Ganoderma resinaceum* (reh zin AY see um) exudes a resinous liquid when cut.

ret / reticulum (L): A net, reticulated. *Peziza reticulata* (reh tic you LAH tuh) is reticulated, as are the over 450 other species with similar epithets. A reticulation is something that resembles a net or network, and the reticulation can be the pattern observed on spores, or it could also be used as an adjective to describe them as on the upper portion of the stem on *Boletus*.

revolut (L): To roll back. Revolutions roll back the time, but species with the epithets *revolutus* (re voe LOO tus) and *revoltum* just have an inrolled margin.

rhac (G): Rags, tattered. What we call *Chlorophyllum rhacodes* (rha COE dees) was initially called *Agaricus rachodes* by Carlo Vittadini in 1835. This species was transferred to *Lepiota rachodes* in 1872 and then *Macrolepiota rachodes* in 1949. Have you noticed that we spell it 'rhacodes' while it was traditionally spelled 'rachodes'? This is all because of a little spelling error in the early 1800's. Whoops. It almost certainly should have been spelled 'rhacodes' since it was named after the tattered appearance of the cap. In 2002, it was put in the genus *Chlorophyllum*, and it is listed under both *C. rhacodes* and *C. rachodes* at mushroomobserver.org. However, it is acceptable according to the International Code of Nomenclature to correct old misspellings and many mycologists agree it should just be '*rhacodes*'.

rheo (G): To flow. *Lactarius chrysorrheus* (chry sor RHE us) has yellow flowing milk. Related words include the Spanish 'rio' for 'river', the 'Rhine', the river in Europe, and 'rhinestone'.

rhod (G): Red. *Russula rhodopus* (RHO doe pus) has a red colored stipe. Also used in rhododendron.

rhododendron (G): Rose-Tree, Rhododendron. *Hypocreopsis rhododendri* (rho doe DEN dry) is an ascomycete that grows on rhododendron and mountain laurel.

rhomb (G): Whirling, a spinning top, a shape represented by four equal sides with two acute and two obtuse angles. Many species have rhomboid-shaped spores, like *Entoloma rhombisporum* (rhom bih SPOR um). I strongly suggest you Google this species and check out a picture of its spores.

rhopal (G): A club. *Amanita rhopalopus* (rho PAL oh pus) is nick-named the 'American club-footed Lepidella'.

rhytism (G): A patch, spot. *Rhytisma* are plant pathogens that cause black spots on leaves and go by the common name 'tar spots'. For example, *R. americanum* is the 'maple tar spot'. A related word is 'rhytid', which is a wrinkle, and 'rhytidoplasty' is a face-lift to remove wrinkles.

ribis (L): After the genus name of the currant plants, *Ribes. Phylloporia ribis* (RYE bis) grows on currant and gooseberry bushes. *Puccinia ribis* is one of a large number of pathogens on the currants.

ric (L): A veil. *Cortinarius riculatus* (ric you LAH tus) has a pronounced veil, as do other species with this and similar epithets.

Ricken, Adalbert (1851 – 1921) was a German Roman Catholic priest and mycologist. He published a few books on European mushrooms. True to the German roots, it should be pronounced (REE ken el uh), but it is more commonly pronounced (rick en EL uh).

rigens (L): Stiff, rigid. *Cortinarius*, *Entoloma*, and *Tylopilus rigens* (RIG ens) are rigid in one way or another. A related phrase is *rigor mortis.*

rigid / rigio (L): Stiff. Members of Russula section Rigidae (RIG ih day) have firm (stiff), compact, dry caps.

rim (L): Fissure, crack. *Inocybe rimosa* (rhi MOE suh) appears cracked with maturity. Rimose can be used as an adjective to describe any mushroom that has a cracked cap. Have you ever seen a mushroom where the margin of the cap was cracked? Use this word the next time you see it.

ringen (L): To open wide, gape. *Panellus ringens* (RIN gens) has fruitbodies that open up like a fan.

Ristich, Samuel S (1915 - 2008): Sam was a founding force behind the Connecticut-Westchester Mycological Association (COMA) and is the namesake of the annual Northeast Mycological Federation (NEMF) foray. *Amanita ristichii* (riss TICH ee eye) is named after him.

rivus (L): A stream, channel, crack, or groove. *Clitocybe rivulosa* (riv you LOW suh) develops cracks as it ages. You can also use the word rivulose as an adjective to describe cracks or irregular crooked lines.

robinia: The genus of the Locust tree, originally named after French gardeners Jean and Vespasien Robin. In mycology, *Perenniporia robiniophila* (rah bean ee AH phil uh) is associated with black locust (Locus trees are in the genus *Robinia*) and hackberry trees. A specimen that grows on the lawn of the Field Museum in Chicago is featured on both Patrick Leacock's and Michael Kuo's website.

robus (L): Firm and hard, robust. *Tricholoma robustum* (rho BUS tum) is strong and robust.

Rodman, James was a Program Coordinator in the National Science Foundation until 2006. *Megacollybia rodmani* (ROD man eye) was named after him "in appreciation of his service to biological systematics." He also has deep ties to the Michigan State University herbarium, and established both the MSU Herbarium Endowment Fund and the James E. Rodman Botany Scholarship Endowment. Controversy exists as to whether the epithet should be 'rodmani' or 'rodmanii'. Although 'rodmani' was technically incorrect since this is a 'modern name' ending in an 'n', changing such a minor mistake may not be warranted. However, calling the mushroom either of these names over *Megacollybia platyphylla* is certainly preferred.

Romagnesi, Henri (1912 - 1999) was a French mycologist who specialized in the Entolomas and Russulas, and published several books and monographs. He named *Tricholoma ustaloides*, while *Coprinopsis* and *Amanita romagnesiana* (rho man ee see AN uh, if we pronounce it in the spirit of his last name) were named after him.

ros (L): Rose. *Cantharellus roseocanus* (rho zay oh CANE us) has a light pinkish cap when young, and it turns pale with age and/or sunlight. It is common in the Rocky Mountains.

rot (L): Wheel. *Marasmius rotula* (rho TOO luh) is commonly called the 'pin-wheel mushroom'. The underside of the mushroom looks like a wheel (think of a wheel on a horse-drawn carriage), with the gills resembling the spokes. A 'rotary' (the traffic circle or the old type of phone) comes from the same root.

rotundus (L): Round. *Ganoderma rotundatum* (rho tun DAY tum) is semi-circular and fairly large. In related terminology, a 'rotunda' is a round room, often covered with a dome. Additionally, 'rotund' is broadly used to describe a round person.

rubell (L): Reddish. Over 300 fungal species have epithets like *rubella* or *rubelliana*. 'Rubella' is also a virus that causes a reddish rash.

rubig / rubigo (L): Rusty or rust-colored. *Hymenochaete rubiginosa* (roo big in OH suh) has a rusty colored surface.

rubr (L): Red. *Lactarius rubrilacteus* (rube rih LAC tee us) has red latex.

rubricund (L): Red. *Phallus rubicundus* (RUBE ih CUN dus) is named after its red color.

ruf / rufus (L): Reddish. *Hymenochaete rufomarginata* (ROO foe mar gin AH tah) has a reddish margin. Additionally, *Microglossum rufum* is reddish. Can you think of another species that has *ruf* in it because of a reddish color?

rupis (L): A rock. *Lycoperdon rupricola* (roo PRIC ola) is described as growing on moss covered boulders.

rus (L): The country. Used in species like *Mycena* and *Arrhenia rustica* (RUS tick uh). 'Rustic' and even 'rural' are related words.

russ (L): Red. Used in *Russula* (ROO suh luh) or (RUSS uh luh) – even though not all of them are red. The type specimen of Section Russula is the easily recognized *Russula emetica.* The specific epithet *russuloides* has been used with 11 different genera. The suffix *–oides* denotes a likeness of form, in these cases, to a *Russula.* Most often, these mushrooms are tinged reddish.

Russell, John Lewis (1808 - 1873) was an American botanist and Unitarian minister from New England. He named *Boletus frostii* (now *Exsudoporus frostii*), while *Aureoboletus russellii* is named after him (*'russellii'* is frequently misspelled: remember that it has three double letters).

rutilus (L): Reddish, ruddy. *Chroogomphus rutilis* (roo til is) is reddish. Similar epithets include *rutilans*, *rutilans*, *rutila*, *rutilescens*, and *rutillum.*

S

sabin (L): Juniper. *Gymnosporangium sabinae* (suh BIN ae) is the 'pear trellis rust', and it uses junipers as the primary host, and pear trees as the secondary host.

sacc (L): Sac. *Geastrum saccatum* (sac CAY tum) is the 'rounded earthstar', and is basically shaped like a sac. Another related and common epithet is *sacciformis.*

sacchar (G): Sugar. *Exidia saccharina* (sac car EE nuh) is sweet. *Hebeloma sacchariolens* has a sweet odor. Saccharin is an artificial sweetener.

salic (L): The willow. *Russula saliceticola* (sal ih cet ih COE luh) grows in association with willow trees. In related words, salicylic acid (related to aspirin) is derived from the bark of willow trees.

salm (L): Salmon, salmon-colored. *Russula salmoneolutea* (sal mon ey oh LOO teh uh) is salmon-yellow in color.

sambuc (L): The Elder tree, genus *Sambucus. Cercospora sambucicola* (sam bu cih COE luh) and *Xylodon sambuci* grow on elder trees.

sangui (L): Blood. *Cortinarius* section Sanguinei (sang GUIN ee eye) are red. The words 'sanguine' (blood-red) and 'sangria' are related.

sanios (L): Full of bloody pus. Fries originally named *Cortinarius saniosus* (san ee OH sus) because of a blood-red milky substance he noted on the lower stipe. Curiously, this feature has not been noted in other collections but the name has stuck. Related to 'sanguine' and 'sangria'.

sapid (L): Having a pleasant taste, Savory. *Boletus sapidus* (SAP id us) and the many other species that use this epithets taste good.

sapine (L): Fir tree. *Russula sapinea* (sap in EE uh) are associated with fir trees.

sapon (L): Soap. *Tricholoma saponaceum* (sah poe NAY cee um) has a definite, albeit mild, soapy odor.

sarc (G): Flesh, fleshy. *Sarcoscypha* (sar co SKY phuh) and *Sarcosphaera* are both fleshy ascomycetes. In related terms, 'sarcasm' literally means 'to strip off the flesh'. Also, a 'sarcophagus' is a stone chamber to rest the dead, and it literally means 'flesh-consuming'. In medical terminology, 'sarcoma' and 'sarcoplasm' are related words.

sardonia (G). Bitter, scornful. *Russula sardonia* (sar doe NEE uh) joins the ranks of *Russula fellea* and *R. astringens* as unpleasant-tasting mushrooms.

saturn (L): A god of agriculture. In some epithets, it refers to the 'gloomy' nature of the planet. *Cortinarius saturninus* is found in gloomy places, *Penicillium saturniforme* is Saturn-shaped, and *Cyberlindera saturnus* has Saturn-shaped ascospores.

sax (L): A stone. *Hygrophorus saxatilis* (sax uh TIL is) literally means 'Hygrophorus among the rocks'. It doesn't always grow around rocks, but I'm assuming it was the first time it was found. The striped bass, also called the rockfish, is *Morone saxatilis*, and the epithet is used several times in the plant world too.

scabr (L): Rough, scurfy. This is a common adjective to describe the so-called 'scaber stalk boletes' of the genus *Leccinum*, which describe the rough appearance of the stem. It is also used in around 380 mushroom names, including *Sarcodon scabripes* and *Lepiota scaberula.*

scalpium (L): Scratched, pick. Used in *Auriscalpium* (aur ee SCAL pee um), the 'ear-pick fungus', and *Arrhenia auriscalpium.*

scamb (G): Bow-legged. *Pholitoa scamba* (SCAM buh) has a curved stipe.

scaur (G): Projecting ankles. The base of *Cortinarius scaurus* (SCAUR us) resembles an ankle.

schiz (G): To split. *Russula schizoderma* (skits oh DER muh) has split skin. It is also used in the word 'schizophrenia', which literally means 'split mind', or 'split personality'. Because of the literal translation here, people have incorrectly thought that schizophrenics have multiple personalities, but 'multiple personality disorder' is a very different disease.

de Schweinitz, Lewis David (1780-1834): Considered to be the father of North American Mycology, it is only fitting that one of the most common polypores posted on identification sites, *Phaeolus schwinitzii* (shwhy NIT zee eye), is named after him. Although he devoted most of his life to the church, in his spare time he compiled descriptions of fungi (including over 300 species that were undescribed at the time).

scler (G): Hard. *Scleroderma* (SCLARE oh DER muh) literally means 'hard skin', and these species are certainly hard, especially compared to their puffball allies.

scob (L): Sawdust. *Coprinopsis* and *Psilocybe scobicola* (sco bih COE luh) grow on sawdust. As an adjective, 'scobiform' refers to a sawdust-like form.

scop (L): A broom, thin branches. *Eutypella* and *Pestalotiopsis scoparia* (sco PAR ee uh) resemble brooms or have broom-like appendages. A note that there was an Italian naturalist in the 18th century named Giovanni Antonio Scopoli. The plant genus *Scoparia* was named after him, and a chemical derived from the plant, 'scopolamine', was named after the plant. Further, there are a few plant pathogens with the epithet *scopariicola*, named because they grow on *Scoparia* species.

scor (G / L): Dung, excrement. *Scorias spongiosa* is the 'Beech Aphid poop-eater'. The Beech blight aphid leaves behind a honeydew (excrement) where the sooty mold fungus grows. As a slight caveat, the genus was named in 1832, and I can't confirm that the fungus wasn't named after its appearance to poop, rather than the aphid substrate it consumes.

scord, scorodon (G): Garlic. *Marasmius scordonius* (scor oh DONE ee us) is one of several *Marasmius* species that smell like garlic.

scrob (L): Ditch, trench. *Lactarius scrobiculatus* (scro bic you LAT us) has a pitted stem. This word element is used in over 100 fungal species.

scut (L): A shield, a flat disc. The famous ascomycete *Scutellinia*, known as the 'eyelash cups', are shaped like little shields. The meaning is used twice in *Scutellinia scutellata* (scu tel LAH tuh) and *Scutellospora scutata.*

scyph (G): Cup. Members of the genus *Sarcoscypha* (sar co SKY phuh) are cup-shaped. Together, the genus name means 'fleshy-cup'.

seb (G): Grease, tallow. *Cortinarius sebaceus* (seh BAY shus) is the color of tallow. The sebaceous gland are small glands under the skin that secrete an oily substance that can make our skin oily. Also, a sebaceous cyst is a relatively common cyst most commonly found on the neck.

secotium (G): Chamber. *Secotium* (seh COE tee um / seh COE shum) is a genus of secotioid fungi – fungi that are forms of gilled or pored mushrooms that look more like a gasteromycete. Thought to be an evolutionary adaptation to dryer conditions, the spores develop within a hidden chamber that is not exposed to the external environment. All secotiaceous fungi used to be placed in the genus *Secotium*, but now they've been relocated to the genus that corresponds to their non-secotioid forms. As a common example, *Secotium pingue* is now *Cortinarius pinguis* (formerly a *Thaxterogaster*). Since so many species were renamed, we have a genus that is mostly gone. Sequestrate is a synonym of secotioid and is frequently used as well.

sector (L): Section of a circle. *Trichaptum sector* represents a portion of a circle. The derivation of this meaning is from 'to cut', think 'to cut a circle', hence the words 'section' and 'sector'.

sejunct (L): Separated. The gills in *Tricholoma sejunctum* appear to separate from the stem (they are just notched, but some describe it as 'deeply notched').

semi (L): Partially. *Gymnopus semihirtipes* (sem ee HER tih peas) has a partially hairy stem.

semital (L): A footpath. *Coprinopsis semitalis* (sem ih TAL is) grows alongside trails, and *Lyophyllum semitale* (sem ee TAL eh) was described from a footpath in coniferous woods. A related phrase is '*alta semita*', meaning 'the high path'.

semper (L): Always. The epithet of *Psilocybe sempervivae* (sem per VEE vay) means 'always living' due to its hallucinogenic properties. The motto of the US Marines, *Semper Fi,* means 'always faithful'.

senilis (L): Old. *Clitocbye senilis* (see nil is) has a wrinkled pileus. 'Senile' is a related word.

senseco (L): To grow old. *Hebeloma senescens* (sen ES cens) is hoary and has a white apex of the stem and generally appears to look old. 'Senile' is a related word.

sensibil (L): Sensible, sensitive. *Boletus sensibilis* (sen sih BIL is) is sensitive to bruising (readily turns blue upon bruising).

separ (L): Divide, separate. The tubes of *Xanthoconium separans* (SEP ar ans) easily separate from the cap.

sepia (L/G): Cuttlefish, a dye from the cuttlefish that produced a red-brown pigment, sepia. *Gloeophyllum sepiarium* (se pee AR ee um) is a brown or reddish-brown color.

sept (L): Seven. *Morchella septimelata* (sep tim eh LAH tuh) was in the seventh clade of Morel species in O'Donnell et al. (2011). The epithet *septemseptata* is a mouthful, but is common and refers to a species having seven-septate ascospores. *Psilocybe septembris* is found in September, and September got its name because it was the seventh month of the Roman calendar. The sixth month was indeed 'sextilis' – and it is probably good that 'sextilis' or 'Sextember' wasn't carried over to the Gregorian calendar. *Septoria,* the fungus that causes leaf spots, is unrelated and named after 'septum', which means wall.

septentrionalis (L): Belonging to the north. *Scerloderma septentrionale* (sep ten trio NAL eh) is similar to *S. meridionale,* but it has a more northern distribution. You may also be familiar with *Climacodon septentrionale,* the 'Northern Tooth Fungus'.

septicus (G): Rotten, putrefying. *Pleurotus septicus* grows on rotten wood, and *Fuligo septica* (SEP tic uh), the 'dog-vomit slime mold', has a putrid smell. You can think of 'septic system', 'sepsis', and 'antiseptic' (against putrefaction) as related words.

sepult (L): Bury. Used in many buried or partially buried species like *Geopora sepulta* (SEP ul tuh). A 'sepulcher' is a small monument where someone is buried.

sequoia (L): The genus of the redwood tree, itself named after Sequoyah, a Cherokee who created the Cherokee writing system. Many species grow in association with the tree go by the epithets *sequoiarum* or *sequoiae* (how many other words do you know that have all five vowels right next to each other?).

serenus (L): Clear, bright. *Leucoagaricus serenus* (ser EE nus) and *Tranzscheliella (=Ustilago) serena* are clear or bright. A related word is 'serene', which means clear and calm.

seri (G): Silk, silky, originally named after the Seres, a group of people from Serica who made the first silk. This is used in over 40 silky species, including *Cortinarius sericellus* (ser ih CELL us), *Lycoperdon sericellum*, and *Alboleptonia sericella*. 'Sericulture' is the rearing of silkworms.

serotinus (L): Becoming ripe late. *Sarcomyxa serotina* (formerly *Panellus serotinus*, the late oyster) and *Russula serotina* (ser oh TEEN us) fruit late.

serp (L): Snake, creeping. *Serpula* is a genus that causes a dry rot to creep through the wood. Related, if you 'serpentine', you move in a zig-zag pattern like how a snake moves.

setos (L): Hairy. *Scutellinia setosa* (seh TOE suh) has thin hairs around the cup. In related words, 'setae' are fine hairs on the body of earthworms that help them move.

sex (L): Six. *Morchella sextelata* (sex tel AH tuh) was in the sixth clade of Morel species elucidated in O'Donnell et al. (2011).

shii (Japanese): Oak tree. The common name *Shiitake* (shi TOK ee) refers to the growth of *Lentinula edodes* on oak.

sibiric (L): Latinized form of Siberia. Over a hundred species with the epithets *sibirica*, *sibiricus*, and *sibiricum* were all described from Siberia. Ironically, *Suillus sibiricus* (sigh BEER ih cus) is the name given to the western North American equivalent to *S. americanus*.

siccus (L): Dry. *Marasmius siccus* (SIC cus) is dry: the stem is dry, the cap is dry, the gills are dry, and you can find them on dry leaves.

sienn (I): A reddish-brown color. *Conocybe siennophylla* has reddish-brown gills.

silv (L): Woods, forest. *Agaricus silvicola* (sil VIC oh luh) grows in the woods and is sometimes called the 'Woodland Agaricus'.

sinap (G, L): Mustard. *Hebeloma sinapizans* smells like mustard and *Pulveroboletus sinapicolor* is colored like mustard.

sinensis (L): Of China. *Elaphocordyceps sinensis* (sih NEN sis) (formerly *Cordyceps sinensis*) is a *Cordyceps* species described from China. *Sinae* is an ancient Greek and Roman word for people in an area now known as China.

sipapu: Named after the Sipapu Ski Resort, which is 20 miles outside of Taos, New Mexico. *Agaricus sipapuensis* (see puh poo EN sis) is a summer Agaricus found near the resort at the New Mexico Mycological Society foray in 2010. This is a new species described in Kerrigan's 'Agaricus of North America'.

smaragd (G): A precious stone that is light green, emerald. *Russula smaragdina* (smar AG din uh) is a pretty light green. There is also an emerald betta fish (*Betta smaragdina*), a weaver ant with an emerald gaster (*Oecophylla smaragdina*), and there is a famous emerald tablet called the *Tabula Smaragdina.*

Smith, Alexander Hanchett (1904-1986): The genus *Smithiomyces* is named after him. He was an avid collector and prolific publisher. He once told the Colorado Mycological Society, "I've been collecting for 46 years and have never seen a 'typical' year for mushrooms". – Or at least that is what they told me, he died when I was 10. Smith is such a common surname that *Cortinarius ahsii* (AH see eye) was named after his initials.

Smith, Worthington George (1835-1917) was an English author and mycologist. *Clitopus smithii* was named after him. However, Smith is one of the most common surnames in the world, and there have certainly been a lot of them in the mycology world as well. Many are species are named after Alexander Hanchett Smith, who was born in 1904, but around 25 *smithii* species predate AH Smith, and some even predate Worthington George Smith.

Snyder, Leon Carlton (1908 - 1987). Snyder's 1938 dissertation from the University of Washington was a 63-page paper "The operculate discomycetes of Western Washington". One of the species he presented there was *Morchella crassistipa*, but Kuo found that the type collection contained two distinct species. In 2012, Kuo named the species *Morchella snyderi* in his honor.

sobol (L): A sprout. *Ophiocordyceps sobolifera* (SO bol IF er uh) sprouts off of cicada nymphs. It is similar to *Cordyceps hesleri,* but has a pale ocher stromata, and is from Cuba, Mexico, and Asia.

soci (L): Belonging to a group. *Coprinus sociatus* (so cee AT us), as well as *Armillaria*, *Cltopilus*, and *Arcangeliella socialis*, all grow in small groups. Related words include 'social' and 'sociology'.

solan (L): A potato, or any plant in the nightshade family. *Amanita solaniolens* (so lan ee OH lens) smells like a potato. Solanales is a taxonomic Order of the nightshades, while *Solanum* is the genus of potatoes.

solitar (L): Growing alone, solitary. *Russula solitarius* (sol ih TAR ee us) grows alone, and you play 'solitary' alone.

solstit (L): Belonging to the Midsummer, the solstice, but literally, a 'stopped sun'. *Entoloma solstitialis*, and many other species with this epithet, refer to when they have been found. This is the antonym of the epithet *brumalis*, which refers to the winter solstice.

sordescens (L): Becoming dirty, vile. *Auricularia sordescens* (sord ES cens) becomes dirty/vile with age, although I do not have experience with it. *Tylopilus sordidus* has a vile odor. In related terminology, the word 'sordid' means filthy or foul.

soror (L): A sister. Used in *Russula sororia* (sor OR ee uh). Species with the *sororia* name are generally named due to their growth pattern (several specimens growing nearby each other). However, I can find no specific reference to this for *Russula sororia.* A 'sorority' is a house of sisters.

Sowerby, James (1757 – 1822) was an English botanic illustrator and publisher. His biggest work was a 36 volume series called "English Botany" that contained nearly 2,600 of his engravings. Specific to mushrooms, his book "Coloured Figures of English Fungi or Mushrooms" was published in a few volumes in the late 1700's. There are many species with epithets *sowerbyi* that pay homage to him and his work.

spadic/spadix (G): Date-brown, nut-brown. This has been used in over 250 species names as a reference to their color, including *Lycoperdon spadiceum* (spad ih SEE um).

sparass (G): To tear to pieces. *Sparassis* (spar ASS is) has a fruitbody that is in many pieces.

spath (L): Spatula. *Arrhenia spathulata* (spath oo LAT uh), and *Dacryopinax spathularia* both look like a spatula. It is also used in the genus *Spathularia* and in over 150 specific epithets.

specios (L): Handsome, showy. *Hygrophorus speciosus* is good looking.

spectabilis (L): Remarkable, spectacular. Although *Gymnopilus spectabilis* (spec TAB il iss) is deprecated and should now be called *G. junonius*, it is still the predominant name in field guides. What an all-around spectacular mushroom indeed.

speir (G): Wrapped, coiled, spiral. The many species with this word element appear coiled. This is unrelated to the word 'spirit', which is from the Latin for 'breath' (think of the 'spirometer' or 'expirate') or 'soul'.

spelunc (G): A cave. *Laboulbenia speluncae* (speh LUN cay) is found in caves. A related word is 'spelunker', someone who explores caves.

sperma (G): Seed (spore). *Russula cyclosperma* (cy clo SPERM uh) has nice round spores.

sphaer (G): Ball, sphere. *Gyromitra sphaerospora* (sphare oh SPOR uh) has spherical spores.

sphagnos (G): Moss. *Gymnopilus sphagnicola* (sphag NIC oh la) grows with moss. *Sphagnum* itself is a genus of well over 100 mosses.

sphinct (G): Tightly bound, a band. Nearly 150 fungal epithets start out with this word element, generally referring to a band of one kind or another. Related are the body's sphincters, which are just rings of muscles.

spilos (G): A spot. *Ophiodothella leucospila* (loo co SPIE luh) has white spots. It is also used in *Pluteus podospileus*, which has spots on the stem.

spin (L): Thorn, spine, backbone. *Spinellus fusiger* is a mold that frequently grows on *Mycena* and *Collybia* species that makes them look like they are covered in tiny spines.

spiss (L): Compact, crowded, thick. The 80 species with the epithets *spissum* (SPISS um) or *spissus* are thick, dense, or compact in some way.

splachn (G): Entrails. *Marasmius splachnoides* (splach NOY dees) has a gut-like stem.

splend (L): Gleaming, shining. *Aleuria spendens* (SPLEN dens) is a gleaming species. A related word is 'splendid', meaning 'magnificent'.

spong (G): Sponge. *Gymnopus spongiosus* (spongy OH sus) is spongy.

spons (L): Promised, a groom. *Amanita sponsus* (SPON sus) is the unblushing groom of *A. novinupta.*

spor (G): Seed, spore. *Elaphomyces cyanosporus* (cye AN oh SPOR us) has dark blue spores.

sporangium (L and G): Combination of spore + vessel. In most slime molds, the plasmodium forms individual spore-containing sporangia. The element 'angium' is used in the fungal genus *Hysterangium*, which literally means 'womb-vessel'. In the plant world, angiosperms produce encased seeds. Although not a fungus, slime molds are often collected by mushroom hunters, so these basic words should be in our lexicon.

Sprague, Charles James (1823-1903) was a New England mycologist in the 1800's. He collected what would become *Suillus spraguei* (SPRAG ee eye) in 1856 and did some exquisite pen and ink illustrations of what he saw. Many other species are named after him as well.

spreta (L): Despised, hated. *Amanita spreta* (SPRET uh) is dubbed the 'hated caesar'. Alan Bessette says that it is not poisonous and posits that it got its reputation because of a similarity to *A. phalloides.*

spum (L): Foamy, spongy. *Pholiota spumosa* (spu MOE suh) is spongy. A related word is 'sputum', the frothy mix of saliva and mucus you cough up on the tail end of a cold.

squal (L): Dirty, filthy, and dry. Used in *Lactarius squalidus* (SQUAL id us) and *Psilocybe squalidella*, which Peck reported was squalid when old. 'Squalid' and 'squalor' are related words.

squamulosus (L): Covered in scales. *Gymnopilus squamulosus* (squam you LOW sus) is scaly, *Gymnopilus fulvosquamulosus* has tawny scales, and *Polyporus squamosus* is also scaly. You can use 'squamulose' as an adjective.

Squarepants, Spongebob (1999 – current): *Spongiforma squarepantsii* was named in 2011. The fungus is spongy, orangish, and bears an uncanny resemblance to Mr. Squarepants, for which it was eponymously named.

St. George's mushroom: The common name of *Calocybe gambosa*, which typically fruits around St. George's day (April 23rd) in the United Kingdom and elsewhere.

stagn (L): Swamp. *Phaegalera stagnina* (stag NI nuh) was originally described from bogs. A related word is 'stagnant'.

Stahel, Gerold (1887 - 1955) was a botanist and mycologist in Surinam, who mostly studied fungi that negatively impacted crops of the area. But also, Stahel discovered an undescribed stinkhorn and sent it to Eduard Fischer. Fischer published his description in 1921 and named the genus *Staheliomyces* after him.

stannum (L): Tin, the color of tin (greyish). *Clitocybe* and *Mycena stannea* (STAN ee uh) are tin-colored. Tin has the abbreviation 'Sn' on the periodic table of the elements.

steinpilz (German): Stein (stone) + pilz (mushroom). 'Steinpilz' is one of the many common names for *Boletus edulis*, and this one refers to the hard flesh of the mushroom. Related, a 'stein' is a large mug for drinking beer.

stell (L): A star. *Scerloderma stellatum* (stel LAT um) opens up at maturity and looks like a star.

stemon (G): A thread. The slime mold *Stemonitis* (stem on I tis) is a mass of thread-like projections.

steppe (Russian): An ecosystem of Eurasia. *Morchella steppicola* (step IH coe luh) was originally described from this region of the Ukraine in the 1940's. Phylogenetically, this morel is the oldest morel species known.

ster (G): Solid, tough. Members of the genus *Stereum* (STAIR ee um) are generally tough. A related word is 'stereo', which means 'hard recording'.

stercoris (L): Dung. *Arrhenia stercoraria* (ster cor AIR ee uh) is associated with dung. 'Sterculius' was a Roman god of manure.

sterquilin (L): A dung-pit, dung-heap, a midden. *Coprinus sterquilinus* (ster QUIL in us), the 'Midden Ink Cap', grows on horse manure. The literal meaning of the full species name is a little redundant, 'dung-living on dung-heaps'.

Stevenson, Rev. John (1836 - 1903) was a Scottish author and mycologist. In 1879 he published 'Mycologia Scotica: the fungi of Scotland and their geographical distribution'. *Collybia stevensonii* is just one of a number of species named after him.

stict (G): punctured, dotted. *Polystictus* (pol ee STICK tus) species appear to have many spots, while *Mycena stictopus* has a spotted stipe. *Sticta* is also the genus name of the 'spotted felt lichens'. 'Stiction' is unrelated, and is a portmanteau of 'static' and 'friction'.

stigm (G): A point or a mark. *Diatrype stigma* (STIG muh), and other species with this epithet, are characterized by spots/points/marks. In flowers, the stigma is the bump at the end of the pistil. More generally, a stigma is a mark, brand, or even a negative mark associated with belonging to a particular group, while stigmata (plural) are the crucifixion marks on the body of Jesus.

stilla / stillatus (L): Drop/many drops. *Chroogomphus stillatus* (stil lat us) has many little droplets. In related terminology, the word 'distill' means 'coming down in drops'.

stiptic (G/L): Astringent. *Panellus stipticus* (STIP tih cus) is sometimes called 'the astringent Panus'. This is not an edible mushroom, though, and McIlvaine didn't even eat it. 'Astringent' itself means 'to contract or pull tight', as the lips do upon exposure to something like this.

stramin (L): Straw, made of straw, or generally straw-colored. *Asterodon strammineus* (stram in EE us) is straw-colored. A much more common species that uses this word element is *Floccularia straminea* (formerly *Armillaria*), which is pale or straw yellow.

strata (L): A layer. *Hypochnicium stratosum* (strah TOE sum) has a layered appearance in cross section.

striatum (L): A furrow or channel, grooved. *Ganoderma crebrostriatum* (CREE bro stri AY tum) has thick/crowded grooves. There is also an area in the brain called the striatum. It is made up of several subnuclei and cell types, and thus appears striated when it is sectioned.

strig (L): Covered in short hairs, can also mean thin, meager. *Lentinus strigosus* (strig OH sus) is densely covered in short hairs.

strobilos (G): A pine cone. *Strobilomyces* (stroe bil oh MY seas) species resemble pine cones.

stromb (G): Top-like. *Gerronema strombodes* has fruitbodies shaped like tops (more-so when young?).

stroph (G): Belt, sword-belt. Members of the genus *Stropharia* (stro FARE ee uh) have a ring around the stem (annulus).

stygos (G): The lower or underworld. *Lyophyllum eustygium* (ee you STY gee um) means 'truly from the underworld'. It was named by Cooke in 1891, perhaps because it becomes completely black with age and has a rancid mealy smell. Or maybe there was another explanation. In words you aren't likely ever encounter, 'stygian' means 'extremely dark and gloomy'.

styl (G): Pillar. The epithet of *Hebeloma ischnostylum* (ish no STY lum) refers to the stipe as a pillar, and 'ischnos' is Greek for 'thin', so together we have a 'thin stipe'. The 'stylus' of a record player or even the small pen-like device used for touch-sensitive monitors are related. Also, the 'styloid process' is a small pillar-like protrusion of the temporal bone.

suav (L): Sweet, agreeable. The epithet in *Hydnellum* and *Inocybe suaveolens* (sue ah veh OH lens) indicates that they have a pleasant odor. I used to throw a *Hydnellum suaveolens* in my car every now and then because they acted as an air freshener - and they even lasted a month or so.

sub (L): Below, almost, near. *Russula subnigricans* (sub NIGH grih cans) is similar to *R. nigricans*. This is a very common prefix. A search of *sub* in Indexfungorum yielded the maximum number of hits the search engine can display (which I found out is 6,000).

suber (L): Cork. Used in over 100 specific epithets, which all reference cork in some way. *Gymnopilus suberis* (sue BEAR iss) lives on cork oak, *Penicillium subericola* was isolated from cork, and still others allude to a cork-like property of the fungus. Also, the 'cork oak' is *Quercus suber.*

subtil (L): Slender, minute, delicate. There are over 250 slender or delicate species that include this root, including *Ramaria subtilis* (sub TIL is), *Mycena subtilipes* (referring specifically to the stem), and *Lachnellula subtillissima* (which includes the suffix meaning 'very'). 'Subtle' originated with this root.

succ (L): Sap, juice. *Peziza succosa* means 'full of juice', and if this fungus was common enough to have a common name, it would be the 'juicy Peziza' because it readily exudes a yellow juice upon handling. 'Succulent' is a related word.

succineus (L): Amber. *Auricularia fuscosuccinea* (FUES co suc sin EE uh) has a dark amber color. There is also a *Collybia succinea* and a Genus of amber snails called *Succinea.*

sudor (L): Sweat. *Mycena sudora* (sue DOR uh) has such a viscid pileus it is said to sweat.

suillus (L): While *suis* is Latin for 'pig' (the genus of pigs is *Sus*), *suillus* means 'pertaining to a pig'. *Suillus* (sue ILL us) got its name because of the greasy nature of the cap. It is also used in *Cortinarius suillus*, which is said to smell like a pig (although I've never had the pleasure).

sulfur (L): Sulfur, yellow-color of sulfur. The 'chicken of the woods', *Laetiporus sulphureus* (sul FYOUR ee us), literally means 'bright yellow pores'.

Sullivant, William Starling (1803-1873): Sullivant settled in what would become Columbus, Ohio, studied grasses, mosses, and became an authority on bryophytes. He was in frequent contact with Moses Ashley Curtis, and both Montagne and Berkeley named species after him. *Boletus* and *Marasmius sullivantii* (sul lih VAN tee i) are two common species named after him.

Sumstine, David Ross (1870 – 1965). Sumstine was a well-published mycologist from Pennsylvania. He collected over 10,000 specimens, including what would become *Meripilius sumstinei.* Although he only named one species himself, *Meripilius*, *Lactarius*, *Poria*, and *Psathyrella sumstinei* are all named after him. Sumstine was an original member of the Mycological Society of America.

surrect (L): To arise, or straight. *Volvariella surrecta* (sur REC tuh) arises from a mass of decaying *Clitocybe nebularis* fruitbodies. 'Resurrect' and 'insurrection' are related words.

sutor (L): Cobbler. *Sutorius* (sue TOR ee us) was erected in 2012 by Roy Halling et al. for *Boletus eximius*, which was originally described by Charles Christopher Frost (The same guy who *Boletus frostii* was named after). 'Sutorius' was named after Frost's profession – he was a cobbler.

syzyg (G): Joining together. *Syzygospora* (siz yg oh SPOR uh) *mycetophila* joins with and parasitizes *Gymnopus dryophilus.*

T

tabacinus (L): Pale brown. *Helvella tabacina* (tab uh SEE nuh) and *Inonotus tabacinus* are two pale-brown species. The root word yields over 175 hits in Indexfungorum.

tabernae (L): Tavern. *Cantharellus tabernensis* (TAB ern NEN sis) was named after a tavern "near the Stennis Space Center Recreation Area frequented by the local work force at the end of the work day" (Feibelman et al., 1996).

tabescens (L): Wasting away. *Armillaria tabescens* (tab ES ens) is the 'ringless honey mushroom'. Did this species get the name because the mushroom itself wastes away easily, or because it wastes away the host?

tabul (L): A table or board, by extension, flat. *Cortinarius tabularis* (tab you LAR is) has a flat pileus. Related words include 'tabular' and even 'tablet'.

tactus (L): Handling or touch. *Russula luteotacta* (loo tay oh TAC tah) turns yellow with handling.

taen (G): Banded, striped. *Russula taeniospora* (TAE nee oh SPORE uh) appears to have striped spores. Many other epithets use some form of *taen*, and these species have a general striped appearance.

taget (L): The genus name of the Marigold flower, *Tagetes. Marasmius tageticolor* was named because it has colors similar to *Tagetes erecta*, the African Marigold.

take (Japanese): A word for mushroom. Used in the common names *Shiitake* (shi TOK ee) (*Lentinula edodes*), maitake (*Grifola frondosa*), matsutake (*Tricholoma magnivelare*), and more.

tardus (L): Slow, late. *Hydnellum tardum* (TAR dum) was one of the later *Hydnellum* species to be discovered. Related words include 'tardy' (late), 'tardive dyskinesia' (movements with a slow onset), and 'tardigrade' (slow walker).

tazza (Italian): Cup. The ascomycete genus *Tarzetta* (tar ZET tuh) means 'little cup'. 'Taza' is cup in Spanish.

telamon (G): Supporting band, lint. *Telamonia* (tel ah MONE ee uh) was named for the lint-like whitish fibers of the veil.

telephon (E): Telephone. *Pyrenochaeta telephoni* was discovered on the screen of a cell phone, something that you might be more likely to clean now.

tenac / tenax (L): Gripping, holding. *Xerocomus tenax* (TEN ax) was named after the "tenacious hold it has on the substratum" (Smith & Thiers, 1971).

tener (L): Soft, tender, delicate. *Puccinia*, *Clavulinopsis*, and *Sphaeria tenella* (ten EL luh) are tender.

tenuis (L): Thin, narrow, slender. *Phallus tenuis* (TEN you iss) is not very well-endowed (it ranges from 3 to 4 inches). It was originally described from Java, Indonesia, although it is apparently rare there. Google a picture of this mushroom to see a thin little phallus.

tephro (G): Ash-colored. *Tephrocybe* (tef RAH so be) species have an ash-colored pileus.

terfez (Arabic): Truffle. *Terfezia* (ter FEZ ee uh) is a genus of truffles found in the desert of the Middle East.

terra (L): Earth, ground. *Gymnopilus terrestris* (ter RES tris) and *G. terricola* grow on the ground. In related terminology, *Terra Firma* is 'solid ground', and 'extra-terrestrials' come from beyond the Earth.

tert (L): Third. *Lycoperdites tertiarius* is one of two fossilized gasteromycetes to have been found. It was found in Mexican amber from the middle of the Tertiary Period (third period), and is about 22-26 million years old.

tesq / tesc (L): Desert, in the broader sense, a wasteland. *Lactarius tesquorum* (tes QUOR um) was originally described from Morocco. *Sagaranella tesquorum* was found on "waste ground".

tessel (L): Little cubes, checkered. This is used in several hundred species names, including *Lycoperdon tessellatum* (tes sell AH tum), where it refers to a broken up fruitbody, and *Hypsizygus tessulatus*, which has a tessellated pattern on the cap (notice the use of the word as an adjective).

testace / testaceum (L): Covered in brick or tile, or covered with a shell. This is a common epithet often used to refer to the color of brick, a brownish-yellow, rather than its general appearance or texture. *Gymnopilus* and *Lactarius testaceus* (tes TAY cee um) are named after their color.

tetra (G): Four. *Laccaria tetraspora*, which literally means 'four spores'. It is also used frequently in the natural sciences. For example, a 'tetrapod' has four feet.

thapsin (G): Yellow. *Russula thapsina* (thap SIN uh) is a nice yellow color.

Thaxter, Roland (1858 - 1932): Thaxter was an American mycologist and professor of biology and botany at Harvard in the late 1800s. He named the well-known *Wynnea Americana*. Also, the genus *Thaxterogaster* (thax TER oh gas ter), which literally means 'Thaxter's stomach', is named after him.

theion/thejus (G): Sulfur-yellow. *Hygrophorus hypothejus* (hypo THAY jus) has yellow gills.

thele (G): A teat, nipple. Used in the genus *Thelephora* (thel EPH or uh), because of the bumpy underside of the fruiting body. Aren't you glad we don't call *T. terrestris* the 'terrestrial nipple' or *T. penicillata* the 'nipple paintbrush'?

theobrom (G): Literally, 'food of the gods', and aptly used as the genus name of the cacao tree, from which we get chocolate. Nearly 100 fungal species utilize this word element in their epithets. For example, *Mycena theobromicola* grows on cacao pods, but the name can also be used to describe a chocolate-color. In related words, 'theobromine' is an alkaloid found in cacao that is structurally and functionally similar to caffeine.

Thiers, Harry D. (1919 – 2000): Harry was a prominent mycologist and educator who published extensively through the course of a long career. He had several publications on boletes and published with Alexander H. Smith, Roy Watling, David Largent, and many others. *Amanita thiersii* (THEERS ee eye) is named after him. He named *Boletus barrowsii* and *B. pulcherrimus*, to name a few of his more common ones. Finally, the genus *Harrya* was named after him as well. The most common example is *Harrya chromapes* (formerly *Tylopilus* and *Leccinum chromapes*).

thraust (G): Brittle. There are many brittle species with epithets like *thrausta* (THROW stuh) or *thraustus*.

thuja (G): Genus name of the Pacific red cedar, *Thuja plicata*. *Agaricus thujae sp. nov.* is associated with this cedar. Although the cedar is found along the coast between northern California and Alaska, northern Idaho, and western Montana, the mushroom is only described from two counties in the state of Washington.

tilia (L): Linden tree. *Tyromyces tiliophila* (til ee ah fill uh) and a wide range of other species, grow in association with the linden tree.

tinct (L): dyed, colored. *Russula purpureotincta* (pur pure ee oh TINC tuh) is purplish.

titub (L): To stagger. *Bolbitius titubans* (tih TWO bans) leans with age as if it was staggering.

tolype (G): A ball of wool. Some members of *Tolypocladium* (tol yp oh CLAD ee um) look like a ball of wool on a branch (-cladium). The 'Tolype moth' looks like a mass of wool itself.

tomentosus (L): Densely matted. *Chroogomphus tomentosus* (toe men TOE sus) has fibrils on the cap and stem. Can you think of another mushroom that has some derivation of 'tomentose' in the name?

tormin (L): Causing colic, dysentery. *Lactarius torminosa* (tor min OH suh), 'the bearded milkcap', and other species with this or a related epithet can cause gastrointestinal upset.

torque (L): Collar, a twisted necklace. *Agaricus bitorquis* (bi TOR quis) has a double collar.

tort (L): Twisted. *Laccaria tortilis* (tor TIL is) is called the 'twisted deceiver' because of the twisting form of the pileus. Related words include 'contort', 'torque', 'torture', and 'tortoise'.

torus (L): A bulge, swelling. *Russula torulosa* (tor you LOW suh) and many others with the epithet, all have a bulge of some sort. In Botany, 'torus' is synonymous with 'receptacle', a bulge in the stem where the flowers grow from.

torv (L): wild. The North American *Cortinarius torvus* (TOR vus) is wild, but what mushroom isn't?

trag (G): Goat. *Cortinarius traganus* (tray GAN us) smells strongly of goats. Related, the 'tragus' is the generally triangular-shaped part of the ear hanging over the ear canal (and is frequently pierced). It is named as such because of a tuft of hair that can grow there, particularly with age, which is reminiscent of a goat's beard.

tram (L): Thin. *Trametes* (truh ME tees) species are notably thin.

translucens (L): Shining through. Sixty nine fungal species are named after being at least partially translucent.

tremulus (L): Quaking (after the quaking Aspen). *Populus tremula* is the name of Quaking Aspen, while *Phellinus tremulae* (TREM you lay) is a very common polypore that grows on them. *Tremella* is a jelly fungus whose genus name literally means 'a little shaky'.

tri (L): Three. *Leucoagaricus tricolor* (TRY color) has three colors.

trich (G): Hairy. *Tricholoma* (tri cuh LOW muh) means 'hairy border', although most Tricholomas do not have one. *Trich* is also used in about 650 epithets, including *trichopus*, which would mean hairy foot or stem.

trigon (G): Triangular. *Arrhenia trigonospora* (tri gon oh SPOR uh) has triangular-shaped spores.

trinit (L): A triad. The epithets *trinitatis* (trin ih tee AT us) and *trinitensis* refer to three parts. Related is the 'Holy Trinity'.

trist (L): Sad, melancholic, dull. *Tricholoma triste* (TREE stay) has dull colors. I presume that the other 39 species that contain this word element are also not sad, but that they just display dull colors as well. The common Spanish word 'triste' also means 'sad'.

trivial (L): An adjective meaning common or ordinary. It is used in *Cortinarius trivialis* (triv ee AL is), as well as *Lactarius trivialis*, which is called the 'Common Lactarius'.

Trogia, Jacob Gabriel (1781-1865): Trogia was a Swiss Botanist and author. He published several mushroom books in the 1840's on Swiss fungi. In 1837, he recognized that fungi have a vegetative stage where they exist as mycelium, and then go through a fruiting stage (J of Mycology, vol 13, p.74, 1907). The genus *Trogia* (TRO gee uh) was named after him in 1835 by Fries. Today, *Trogia venenata* is best known for being responsible for the 'Yunnan Sudden Death Syndrome'.

trogle (G): Hole, Cave. Used in the epithets *troglodytes* (trog low DIE tees) and *trogrlophilus* to indicate an affinity to holes or caves. A 'troglobite' is an animal that lives in caves, and 'troglophile' describes an animal that can live their whole life in a cave.

truffle (French): Tuber, which is originally from the Latin meaning tumor or knob. 'Truffle' is the common name of members of the *Tuber* genus, a group of highly sought after hypogeous ascomycetes.

trull (L): A ladle. *Ganoderma trulliforme* is named for its likeness to a ladle. Compare with the epithet *trullisata*, which refers to another tool, the trowel.

trullisat (L): Plastered, troweled. *Laccaria trullisata* (trul lis AH tuh) grows in sandy soil, and the caps are often covered in sand, giving the fruiting bodies a plastered look.

truncus (L): A trunk, stem, or truncated. Used in some form or another in over 350 fungus species, like *Boletus truncatus* (trun CAY tus), which has truncated spores.

tsuga (Japanese): Hemlock. *Ganoderma tsugae* (tuh SOO gay) is commonly found on Eastern Hemlock, while G. *lucidum* grows on hardwoods.

tuba (L): Trumpet. The epithet of *Craterellus tubaeformis* (TWO bay FOR mis) (formerly *Cantharellus tubaeformis*) means 'shaped like a trumpet'. This mushroom goes by the common names 'winter chanterelle', 'yellowfoot', and 'trumpet chanterelle'.

tubae (L): Tube. *Craterellus tubaeformis* (TOO bay FOR mis), considered a choice edible, is a Cantharellaceae. This one has a stem that is shaped like a tube. As far as common names, it is called the 'yellow-foot chanterelle' or 'winter chanterelle'.

tuberculum / tuber (L): A tumor, knob, hump. *Ganoderma tuberculosum* (two ber cue LOW sum) is rather knobby.

Tulasne, Louise Rene Etienne 'Edmond' (1815 – 1885): Tulasne made the first monograph on the Nidulariaceae and published a paper on the life cycle of Ergot. Several fungi are named after him, including *Tulasnella*, and the species *Dacrymyces tulasnei* (two LAZ nee eye) and *Ustilago tulasnei.*

Tulloss, Rodham (Rod) E. (1944 - current) is the preeminent authority on Amanitas, and the founder of Amanitaceae.org (and co-editor, along with Dr. Yang). He has been studying Amanitas since 1978 and was mentored by Dr. Cornelis Bas (1928 – 2013). Rod is well-published, a frequent member of NEMF and NAMA events, and has been awarded several honors (including the NEMF eximium award). He was featured on a blog at fungiflora.com called "6 Mycologists you should know". *Amanita tullossiana* (tul loss ee AN uh) was named after him in 2018 to recognize his contributions to mycology.

tumid (L): Swollen. *Tricholoma tumidum* (TUE mid um) has a swollen stipe. 'Tumid' is an obscure adjective for 'swollen', and is used throughout the animal kingdom.

turb (L): Something that whirls, like a top. *Turbinatus* indicates something is shaped like a cone or a top. Indeed, the base of the stipe in *Cortinarius turbinatus* (ter bin AY tus) is shaped like an old spinning top. Related words include 'turbo' and 'turbine'.

turg (L): Swollen. Several hundred species use this word element, like *Cortinarius turgidus* (TUR gid us), whose stem is swollen at the base. Other epithets include *turgida, turgescens,* and *turgidula.* In English, 'turgid' means 'swollen', and 'turgescence' is the process of becoming swollen.

turm (L): Troop. *Cortinarius turmalis* (tur MAL is) grows in troops.

turpis (L): Ugly. *Lactarius turpis* (TUR pis) isn't the prettiest mushroom. 'Turpid', meaning 'foul or unclear', is derived from this word.

tyl (G): A knob, swollen. The name of the genus *Tylostoma* (tie low STOW muh) literally means 'knob stomach', and *Tylopilus*, 'knob cap'.

typhrasa (completely made up): An anagram of *Psathyra. Typhrasa* was erected in 2015 to encompass former *Psathyra* species. They get 10 points for being clever, but -20 points for not coming up with a useful name that gives any indication of a key feature or anything else potentially useful.

typhula (L): A reference to the genus of reed-mace *Typha* (which includes cattails). Members of the genus *Typhula* look like little *Typha*, thus the use of the diminutive suffix –ula. *Psathyrella typhae* fruits on dead Typha stems, and *Mollisia typhae* grows on dead Typha leaves.

tyr (G): Cheese. *Tyromyces* is literally the 'cheese mushroom', while tyrosine is an amino acid found in cheese.

U

uda (L): Moist. Species with this short epithet typically have a moist pileus, as in *Bogbodia uda* (ooh duh).

ula / ulus / ulum / ule (L): A diminutive suffix. The epithet of *Marasmius rotula* (rho TOO luh) means 'little wheel' – appropriate given the size of this small species.

ulig (L): Moisture, growing in a marsh or swamp. Species that use the epithets *uliginosa* (you lije in OH suh), *uliginicola*, and *uliginosus* all grow in this particular habitat. 'Uliginose' means swampy or muddy, while 'uliginous' is an adjective that can be used to describe something that grows in a swampy area.

ulmus (L): Elm tree, genus *Ulmus. Hypsizygus ulmarius* (ul MARE ee us) is called "the elm oyster".

umbilic (L): Navel. *Hydnum umbilicatum* (um bill ih CAY tum) has a small depression in the middle of the cap that resembles a navel. There are over 100 species names that use *umbilicate*, *umbilicatus*, or *umbilicatum*. Some sources spell 'umbilicate' with a double 'l's instead of one, and I am not sure why this disparity exists.

umbo (L): A Shield. Used in *Cantharellula umbonata* (um bon AH tah). Have you ever seen a mushroom with nipple-like protrusion on top? Well, instead of saying it has a nipple, you can say that that shape of the cap is 'umbonate'. Alternatively, you can say that the cap has an 'umbo'.

umbrin (L): Darkened, shaded. *Clavulinopsis umbrina* (um BRY nuh) is dark. It is also used in a large number of deprecated varieties, like *Cantharellus cibarius var. umbrinus*, generally denoting a dark variety of the mushroom. The word 'umbra' is commonly used in astronomy, and an 'umbrella' is frequently used by people who don't want to get wet.

uncia (L): A Roman unit of measurement: an inch. *Clavaria* (*Typhula*) and *Russula uncialis* (un she AL is) are two examples named after their average length of approximately one inch.

undat (L): A wave, undulate. *Rhizomarasmius undatus* (un DAT us) and *Entoloma undatum* are wrinkled, or undulate. The epithet of the dragonfruit is *Undatus*, a reference to its wavy structure.

ungul (L): Hoof, claw. *Inonotus ungulatus* (UN gue LAH tus) is shaped like a hoof. In related terminology, the *ungulates* are the hoofed mammals, including deer, horses, camels, and many others.

ur / oura (G): Tail. *Lysurus* has loosened lobes that stick up into the air like little tails. Related, a 'urite' is a segment of an arthropod, and 'uropod' refers to the last pairs of appendages in a crustacean.

uran (G): The heavenly one. *Mycena urania* (your AY nee uh) is as pretty as a fungus gets. 'Uranium' was named after the newly discovered Uranus for no other reason than to pay homage to the discovery.

urb (L): City. *Inocybe urbana* (ur BAN uh) and *Cortinarius urbicus* can be found in urban areas.

uren (L): Stinging, burning. *Marasmius urens* (UR ens) has a burning, acrid taste.

ursa (L): Bear. *Lentinellus ursinus* (err SIN us) is called the 'bear Lentinus' because of the bear-like color and texture of the pileus. Related, 'Ursa Minor', commonly called the 'Little Dipper', is the 'Little Bear' constellation.

urtic (L): *Urtica* is the genus of nettles, itself named after the Latin 'to burn'. *Coprinopsis urticicola* (ur tic ih COE luh) is named after an association with nettle, and *Puccinia urticate* is a rust fungus on nettle leaves.

ustal (L): Burnt, referring to color. *Mycena ustalis* (oo STAL is) and *Tricholoma ustale* are both a burnt-brown color.

ustilag (L): Burnt, smut, soot. *Ustilago* (oo stil AG oh) *maydis* is commonly called the 'corn smut fungus' because of the appearance it gives in parasitized corn.

uvid (L): Wet, damp, humid. *Lactarius uvidus* (OOH vid us) is known as the 'Damp Lactarius', and refers to growth in damp environments.

V

vacc (L): Cow. *Tricholoma vaccinum* (most commonly pronounced vac SIGH num) has a reddish-brown cow-colored cap. This word element is also used in the word 'vaccine', because of the use of cows to prevent smallpox.

vag (L): Wandering. Used in dozens of species like *Xenasmatella vaga* (VAY guh) (formerly *Phlebia* and *Phlebiella vaga*). Related words include 'vagrant', 'vagabond', and the 'vagus nerve', all of which wander in some way or another.

validus (L): Strong, robust. *Gymnopilus validipes* (val ID ih peas) has a robust stipe. In related terminology, an experimental study should be valid.

vanduzer: Latinized form of the 'van Duzer Forest' in northwest Oregon. *Cortinarius vanduzerensis* (van doos er EN sis) was originally described from this area in 1972.

variegat (L): Variegated, to have different colors. *Coprinopsis variegata* (var eye eh GAT uh), and other species with this epithet present different colors.

veget (L): Lively, vigorous. Species with the epithets *vegetus* (VEG eh tus) or *vegetum* exhibit vigorous growth. The word 'vegetable' is related, but it is ironic that someone in a 'persistent vegetative state' is everything but lively.

velutin / velutinus (L): Velvety. *Coltricia velutina* (vel ooh TINE uh) and *Gymnopilus velutinus* are both are velvety.

venen (L): Poison. *Trogia venenata* (ven en AH tuh) is responsible for Yunnan Sudden Death Syndrome and is sometimes called the world's deadliest mushroom. Related words include 'venom' and 'venomous'.

venet (L): sea-colored. *Cortinarius venetus* is a greenish yellow.

venos (L): Vein. *Disciotis venosa* appears to have thick veins on it.

ventricose (E): Bulbous, swollen, often to one side. Heslter (1969) says that *Gymnopilus ventricosus* (ven trih COE sus) has a conspicuously ventricose stipe. Indeed, the stipe is swollen in the middle. Ventricose is apparently a combination of *ventricle* and *–ose*, and this specific combination was first used in the English language.

vermis (L): A worm. The epithet *vermicularis* (ver mic you LAR is) indicates resemblance to a worm, including a few species named *vermicularispora* indicating worm-shaped spores. *Boletus vermiculosus* was named because it becomes full of worms at maturity (I can't imagine how bad it must have been to get this epithet). This word element has been used in over 200 fungal epithets for various reasons.

vernic (L): Varnished. *Lepista* and *Daldinia vernicosa* (ver nih CO suh), as well as *Cortinarius vernicosum*, appear varnished.

verpa (L): Rod, erection. Used in the genus *Verpa* (VER puh) for obvious reasons. 'Verpa' was also a Latin obscenity for penis.

verruca (L): A wart. *Scleroderma verrucosum* (ver roo COH sum) is covered with scaly warts. With the suffix *–osum* or *–osus* (meaning an abundance), the full species name *verrucosum* means 'full of warts'.

versi (L): Many. *Trametes versicolor* (VER zih color), commonly called the 'Turkey Tail', can display an impressive number of colors.

vesc (L): To feed, edible. *Russula vesca* (VES cuh) is not just edible, but found in large enough quantities to make it worthwhile.

vesica (L): Bladder. *Tremella vesicaria* (ves ih CAR ee uh) is named after the bladder-like swellings. In related terminology, the four vesicles of the brain are fluid-filled reserves of cerebrospinal fluid.

vesper (L): The evening, where the sun sets. *Helvella vespertina* (ves pear TEEN uh) has a western distribution. It is also used in *Cortinarius vespertinus*, which has dull, evening-like colors.

vespid (L): Wasp. *Candida vespimorsuum* (a wasp's sting) got its name because the people who discovered the fungus got stung by wasps. This epithet would be useful if one got stung by wasps at least most of the time the species was collected, but no. Just no.

vestis (L): Clothes, covering, cloak. *Cortinarius vestipes* (VES tih pees), *Rhizopogon vestitus*, and *Mycena vestita* are covered or appear to be covered in some way. In related terms (and no big surprise), 'vest' and 'vestibule' are derived from this word element.

via / vialis (L): *Via* means 'road' or 'way' and *vialis* means 'belonging to the road'. *Gymnopilus vialis* (vie AL iss) grows by (or was originally described by) roads.

viet (L): Shrunken, withered. *Lactarius vietus* was named after its appearance. Unrelated but just as common, species with the epithet *vietnamensis* were discovered in Vietnam.

vilis (L): Cheap, worthless, insignificant. What better way to call a mushroom mediocre and boring than to give it the epithet *vilis* (vil is)? How about calling it 'less than worthless' like in *Entoloma subvile.*

vill (L): Hairy, shaggy. *Crepidotus villosus* (vill OH sus) is hairy, as are many other species with the epithet.

vin (L): Wine. Many species with the epithet *vinosa* (vie NO suh) are wine-colored. Related words include 'viticulture' (the study of grapes for wine-making) and 'vineyard'.

viola (L): Violet. *Cortinarius violaceus* (vio LAY shush) is a stunning violet colored mushroom.

virg (L): Pure. Species like *Hygrophorus virgineus* (vir GIN ee us) and *Lachnum virgineum* are pure white, while species like *Morchella virginiana* are named after the state of Virginia, itself named for Queen Elizabeth, the 'Virgin Queen'.

virid (L): Green (color). *Elaphomyces viridiseptum* (veer id ih SEP tum) has some hyphae that stain green. 'Virid', and similar words, appears many times throughout the fungal kingdom: There are over 1,000 hits for *virid* on indexfungorum!

viros (L): Poison, fetid, muddy, covered in slime. *Amanita virosa* (veer OH suh) (the destroying angel) is deadly, and the recently described *Cantharocybe virosa* (formerly named *Megacollybia virosa*) results in significant gastrointestinal upset.

viscosus (L): Sticky. *Gymnopilus viscidissimus* (vis cid ISS im us) is viscid, or sticky. Coupled with the suffix *issimus*, which means 'very much', this mushroom is very viscid. Interestingly, 'viscous' comes from the Latin *viscum*, which means 'bird-lime'. Birdlime is a sticky substance that trappers and bird traffickers have used to put on branches to trap birds. The birds would stick to the substance like flies on flypaper. Fortunately, the use of birdlime is mostly a thing of the past.

vitell (L): Yolk, egg yolk colored. Used in *Neolectra vitellina* (vih tell EE nuh), and many others, to highlight their color.

vitr (L): Brittle like glass. *Mycena vitrea* (vie TREE uh) is brittle. Related, the 'vitreous humor' (the gelatinous fluid in the eyeball) was named for its glassy appearance.

vitt (L): Ribbon, banded, striped. *Galerina vittiformis* (vit tih FOR mis) has a two-toned cap, as if it is banded.

vola (L): The palm of the hand. *Lactarius volemus* (voe LEE mus) exudes copious amounts of milk, enough to fill the palm of the hand. It was named by Fries in 1821.

vor (L): To devour. *Batrachochytrium salamandrivorans* (sal ah man drih VOR ans) is a chytrid fungus that causes skin infections and death to salamanders. Related words include 'voracious', 'voracity', 'herbivore', and 'fungivore'.

vulcan (L): The fire god. Used in *Geopyxis vulcanalis* (vul can AL is), which, despite the name, does not fruit on burnt ground.

vulgaris (L): Common. *Rhizopogon* and *Morchella vulgaris* (vul GAR iss) are common.

vulpes (L): A fox, genus *Vulpes*. Epithets are usually used to indicate a fox-like color. However, *Vulpicida* is a lichenized fungus, and was believed by locals to kill foxes. 'Velpecula' is a fox-shaped constellation.

W

Wakefield, Elsie M (1886-1972) was an English Mycologist. CG Lloyd featured a full page portrait and biography of one mycologist in nearly every issue of his journal *Mycological Notes*, which he self-published in 75 issues from 1898–1925. It wasn't until June 1924 that he featured his first woman mycologist: Elsie Wakefield. Her name lives on in *Oligoporus*, *Thelephora*, and *Poria wakefieldiae*.

Wasson, R Gordon (1898-1986): It is no surprise that the father of Ethnomycology has a couple of Psilocybin-containing mushrooms named after him, including *Psilocybe wassoniorum* (wass on ee OR um).

Weaver, Evelyn M "Peg" (1919-2015) was a published amateur mycologist from the upper Midwest who specialized in Minnesota Boletes. She is also an author on of *Crinipellis cremoricolor* with Schaffer, and *Psathyrella rhodospora* with Smith. Around 2,500 of her specimens now reside at the University of Minnesota.

Whetsone, Mary S: Whetstone was the founder of the Minnesota Mycological Society (MMS) in 1898. She was also the secretary in the club the early 1900's when she sent CG Lloyd an undescribed gasteromycete. He put it in the new genus *Whetstonia* and named it *W. strobiliformis* (Mycological

Notes, Volume 2, number 22, July 1906). *Amanita whetstoneae* (wet STOW knee ay) is also named after her.

X

xanth (G): Yellow. *Agaricus xanthodermus* (xan tho DER mus) is a yellow staining *Agaricus*. The staining occurs naturally when the base of the stem is scratched or when it is cut in half. The surface of the cap turns yellow with potassium hydroxide (KOH), too. Do not eat this *Agaricus*!

xeno (G): Strange, different. *Strictis xenospora* (zee no SPOR uh) was originally described as having strange spores that "resemble ghosts". (Mycotaxon Vol 5 no 1 p 260). Related words include 'xenograft', a graft from another species, and 'xenophobic', being afraid of a group of people who are different from you.

xeros (G): Dry. *Russula xerampelina* (zer am peh LINE ah) is a common shrimp-smelling Russula that is named after the dry appearance of the cap.

xyl (G): Wood. *Russula xylophila* (zye LAH fill uh) loves wood. A related and very common word is the botanic term 'xylem'.

Y

Yates, Harry Stanley (1888 - 1938). Yates got his Masters and PhD from the University of California at Berkeley. As a student, he collected around 8,000 specimens that would be donated and housed at the University of California Herbarium. Yates collected what would become *Tricholoma yatesii* (originally *Melanoleuca yatesii*) on January 24th, 1913 at the University of California at Berkeley, California. Yates published many works, including the book "The comparative history of certain California Boletaceae".

Z

Zeller, Sanford Myron (1885-1948) was an American mycologist who spent most of his career based at Oregon State University. He specialized in the Gasteroids, and published descriptions of over 140 species. *Tricholoma*, *Russula*, and *Xerocomellus zelleri* (ZEL ler eye) named after him (Zeller found the first specimen of what we call 'Zeller's Bolete' while hunting with Murrill).

Zollinger, Heinrich (1818-1859): *Clavaria zollingeri* was named after him, as were species of plants, seaweeds, and other fungi. He also published in geology, meteorology and mollusks before succumbing to malaria at a relatively early age.

zonatus (L): Banded, zones. *Ganoderma densizonatum* (den si zoe NAY tum) has dense bands, or zones. *G. zonatum* also has pronounced bands.

zyg (G): Joined, yoked. Along with *hyps* meaning 'high', *Hypsizygus* (hip sih ZYE gus) means that it is joined to a tree high up in the tree. A related word is 'zygote', a cell formed by the joining of a sperm and an egg.

5. REFERENCES

Arora D. (1986). Mushrooms demystified. Ten Speed Press.

Bessette A.E., Roody W.C., and Bessette A.R. (2000). North American Boletes. Syracuse, New York: Syracuse University Press.

Beug M.W., Bessette A.E., and Bessette A.R. (2014). Ascomycete fungi of North America: a mushroom reference guide. University of Texas Press.

Borror D.J. (1960). Dictionary of Word Roots and Combining Forms, Mayfield Pub., 1st edition.

Dawson J. (2013). What's in a name? Clathrus archerii. NJMA News, v 43(2), p.8-9.

Emmett E. (2003). Microscopical techniques staining with Congo Red. Field Mycology 4(2); 72-3.

Evenson, V. S. (1997). Mushrooms of Colorado and the southern Rocky Mountains. Denver: Denver Botanic Press.

Feibelman T.P., Bennett J.W., and Cibula W.G. (1996). Cantharellus tabernensis: A new species from the southeastern United States. *Mycologia 88*(2): 295-301.

Fiedziukiewicz, M. (2013). Mushroom Toxins – The Meixner Test (Unpublished thesis). The University of York, Department of Chemistry, United Kingdom.

Foltz M.J., Perez K.E., and Volk T.J. (2013). Molecular phylogeny and morphology reveal three new species of Cantharellus within 20 m of one another in western Wisconsin, USA. *Mycologia 105*(2), 447-461.

Gledhill D. (2008). The names of plants, Cambridge University Press. 4th edition.

Hard, M. E. (1908). *The Mushroom, Edible and Otherwise: Its Habitat and Its Time of Growth; with Photographic Illustrations of Nearly All the Common Species...* Ohio Library Company.

Hesler L.R. (1969). North American species of Gymnopilus. Hafner Publishing Company, New York & London.

Jaeger E.C. (1955). A Source-Book of Biological Names and Terms. Charles C Thomas Pub., 3rd edition.

Kerrigan R.W. (1989). Studies in Agaricus IV: new specie from Colorado. Mycotaxon 34(1): 119-128.

Kuo M. (2008, October). Calvatia booniana. Retrieved from the MushroomExpert.Com Web site: http://www.mushroomexpert.com/calvatia_booniana.html

Leonard L.M. (2006). Melzer's, Lugol's or Iodine for identification of white-spores Agaricales? McIlvainia 16(1), 43-51.

Metzler S. and Metzler V. (1992). *Texas mushrooms.* Japan: University of Texas Press.

O'Donnell K, Rooney AP, Mills GL, Kuo M, Weber NS, Rehner SA. 2011. Phylogeny and historical biogeography of true morels (Morchella) reveals an early Cretaceous origin and high continental endemism and provincialism in the Holarctic. Fungal Genet Biol 48:252–265.

Parmasto E. and Voitk A. (2010). Why do mushrooms weep? Fungi 3(4), 15-17.

Rea, C. (1922). British Basidiomycetae A handbook to the larger British Fungi. Cambridge at the University Press. London.

Ryvarden L., and Johansen I. (1980). A preliminary polypore flora of East Africa. Fungiflora, Oslo.

Singer R. (1986). The Agaricales in Modern Taxonomy (4th ed.). Königstein im Taunus, Germany: Koeltz Scientific Books

Stevenson J. British Fungi (Hymenomycetes) Vol. II. Cortinarius – Dacrymyces. William Blackwood and Sons. Edinburgh and London. 1886.

Sung G-H. and Spatafora J.W. (2004). Cordyceps cardinalis sp. Nov., a new species of Cordyceps with an east Asian-eastern North American distribution. Mycologia 96(3), 658-666.

Tortelli M. (2004). The use of Guaiac in the identification of Russula. Field Mycology 5(3).

Made in the USA
Middletown, DE
04 August 2020

14507579R00089